ADAPTING GENETICS

Quantal Evolution After Natural Selection

Surviving The Changes To Come

Roger L Dawkins

NU Publishing

Published by Nearurban LLC, 1603 LBJ Freeway,
Suite 280, Dallas, TX 75234
Tel (972) 974 7130, Fax (214) 393-4001
Email : admin@nearurban.net

First Edition – 2015 - 2nd Impression
ISBN 978-0-9864115-1-9

Dawkins, Roger, L. 1941 –
Adapting Genetics
1. Speciation. 2 Hereditary. 3 Genomics.
1. Title
Manufactured and printed by CreateSpace.

About the title:
Genetics and evolution, as taught for a century, imply that survival is a lottery dependent upon random errors in copying the genome. ***Quantal evolution*** refers to the partitioning of the genome to protect critical regulatory DNA sequences, including during the successive ***retroviral*** epidemics which defined human ***ancestors***. Trial and error selected pre-tested combinations which are faithfully inherited thereby allowing the new species to adapt by changing gears in the ***absence*** of natural selection or mutations. Heterozygous ***polymorphisms*** interact polymorphically, thereby creating additional options. There are clear implications for ***forensics***, ***livestock selection*** and ***transplantation*** inter alia. Anamnestic evolution ensures individuality, diversity, protection from repeat epidemics and survival in the future.

" *I enjoyed the book very much. It is a refreshing look at genetic evolution, which is really needed.*

Professor Stephen Oppenheimer, author of "Out of Eden" and other books.

" *I could not stop reading it. How good it is to see that Professor Dawkins has not changed his enjoyable way of arguing.*"

Professor Jon van Rood, leading HLA pioneer whose work helped establish the clinical and scientific study of the Major Histocompatibility Complex.

Contents

About the Book and its Author

Professor R. L. Dawkins (MD, DSc, FRACP, FRCPA, FRCP) is Chairman and Research Director of the C.Y. O'Connor ERADE Village Foundation (CYOEVF), Emeritus Professor in the Faculty of Medicine and Dentistry at the University of Western Australia (UWA) and Adjunct Professor in the School of Veterinary and Biomedical Sciences at Murdoch University, Western Australia.

From 1973-1997 Dawkins was Founder and Head of Clinical Immunology at the two major teaching hospitals in Western Australia (Royal Perth Hospital and Sir Charles Gairdner Hospital). Within this framework he developed clinical services, diagnostic and research laboratories, the curriculum for the Faculty at UWA and established research in the fields of clinical immunology, rheumatology, transplantation and genetics.

In addition, Dawkins became the Founder, Chairman and Research Director of the Immunogenetics Research Foundation (IRF) from 1983, Director of the Institute for Molecular Genetics and Immunology from 1993 and UWA Director of the Centre for Molecular Immunology and Instrumentation from 1994. He progressively relinquished hospital responsibilities and concentrated on building an international collaborative research group pursuing fundamental issues in genomics, evolution and disease. As its research field expanded, the IRF and Institute were incorporated by the CYOEVF in 2002.

Dawkins is an acknowledged international authority on immunogenetics and autoimmune disease as reflected by; independent funding and awards including the Humboldt Research Award and Program Grants from the National Health and Medical Research Council, appointments including President of the Australian Society for Medical Research, External Examiner for PhD and MD theses, Visiting Professor at the University of Leiden in

The Netherlands, the Max-Planck-Institute in Germany and the Tongji Medical University in China, contributions to national and international councils, committees, symposiums, congresses, chapters, reviews and editorial boards, requests to help establish tissue typing in Thailand and China and, importantly, the success of his former trainee physicians and PhD students.

The work of Dawkins and his group has implications, both practical and theoretical, for the way we view biomedicine, selective breeding of livestock, evolution and inheritance. Their study of the major histocompatibility complex (MHC) showed that there is an alternative way to understand genomic polymorphism. The mainstream view was, and still is, that MHC polymorphism in the population maintains robust antigen-presentation modes and T cell diversity. On the other hand, the Dawkins group showed that DNA sequence diversity and thus polymorphism, extend through the entire MHC irrespective of function. As this book will show, these concepts come together as the *"Multi-functional Polymorphic Controller",* which facilitates both the creation and regulation of genome diversity. From this perspective the presentation of antigenic peptides for recognition by T cells is just one readily accessible marker of MHC polymorphism.

There are three main themes: *"Genomic Lessons from the MHC", "Selective Breeding of Livestock"* and *"The Why and How of Evolved Individuality"*. These essays carry a set of profound and pragmatic messages for the biomedical and veterinary communities. Indeed, Dawkins brings to this task over forty years of practical experience as a physician, clinical immunologist and medical researcher as well as an expert cattle breeder and principal of a paddock to plate mittelstand. He is therefore in a unique scientific position which has led him to his evolutionary analyses and his measured critique of mainstream evolutionary thinking. It also leads to a new concept of inherited individuality.

The essays are written in a warm, engaging and erudite style and are of the type that will stand the test of time.

Edward J Steele, Editor, Nearurban Publishing, 9 October, 2014

Preface

In spite of the prevailing wisdom at the time, we discovered that the "really important" defining DNA sequences have been actively conserved over thousands of generations from the time of the original humans at least 100,000 years ago. By "really important" we mean critical in determining the differences between individuals and therefore the capacity to adapt, compete and survive. Individuality is inherited; it is not created *de novo*. Each individual has an absolutely unique combination of ancestral elements. At each generation, the parents contribute their carefully preserved building blocks but these are shuffled and remixed to generate renewal. How could this be?

Here we address this question and show that the answers are of real importance to medical practice, livestock breeding, forensic investigation, legal ownership, the popular concept of individuality and even terrorism.

It becomes clear that current humans are well prepared for future challenges, such as epidemics, climate change, food shortages, etc and are entitled to adopt a more optimistic view than prevails today.

The above discovery, dating to the early 1980s, was not expected and met fierce opposition. Amazingly, it was, and still is, believed that there is a standard human sequence and that differences are created by continuing mutation through errors in copying and then selected by Darwinian evolution, resulting in the "survival of the fittest". This conviction is patently wrong but survives in spite of the certain fact that mutation would disrupt the inheritance of the advantages gained by earlier ancestors.

We have been surprised by the strength of entrenched convictions and the resistance to change. One conclusion is that commercial forces are intent on selling simplistic turn-key tests so as to recover their investments even though based on outdated concepts. In truth, their financial success provides them with an opportunity to be totally pragmatic and to adopt the current understanding of the genome as it really is. The challenge is to replace the old with the new and to accelerate the use of technology allowing practical benefits to all. At stake is improved health, food and security, to list just a few universal concerns addressed in the chapters below.

This book is intended to expose some of the impediments to adopting a more optimistic future.

The Glossary which follows is important because the terms are crucial but often used idiosyncratically depending upon persuasion.

R. L. Dawkins, August 2014

Acknowledgements

The author acknowledges the contributions of his CYO Foundation Faculty colleagues, Dr Sally S. Lloyd and Associate Professor Joseph F. Williamson for assistance in many aspects of writing these essays, and to Sue Lester for her help in genomic analyses over many years and her critical pieces on Estimated Breeding Values. Mr Dom Bayard of Global Reproductive Solutions is acknowledged for his practical insight and knowledge, particularly on the Japanese Black cattle Wagyu and their main sires in Australia. I thank Dr S. T. Dawkins for the human figures with meshing gears. I also thank Mr Jack Reddin for access to the memoirs of W. J. Dawkins who developed the Poll Dorset sheep in South Australia (1917-1960). Acknowledgements for other animal photos or identification included in the text. Where identification is not indicated these photos were taken by the author of his own cattle.

Further acknowledgements are given in Chapters 2, 3 and 4.

Glossary – Advice on usage of potentially confusing terms.

Anamnestic evolution - relating to memory - reuse of the tried and tested, including in new combinations.

Ancestral haplotype - an alternative sequence of a PFB inherited over generations without recombination; the haplotype equivalent of alleles at a locus. Also known as extended, population or conserved haplotypes.

cis - on the same chromosome

Dominant - independent expression of either maternal or paternal allele or haplotype.

Duplicon - largely repeated sequence attributed to duplication of an earlier sequence. Also used for multiple duplicons.

Epistasis - interactions beyond *cis* and *trans,* e.g. between different chromosomes.

Genomics - the architecture of the genome.

Heterozygous advantage - the benefit of possessing two or more different alleles or haplotypes.

HLA-A,B,C… - loci defined by WHO Nomenclature Committee.

Indel - insertion or deletion defined by comparing sequence with dot plots.

Linkage Disequilibrium (LD) - the combination of alleles which occurs more commonly than expected by multiplying the individual population frequencies. LD does not identify ancestral haplotypes.

Meshing - the additional outcomes generated by alternative engagements of interacting polymorphisms, as illustrated by engaging different cogs in a gearbox. May be stochastic or reactive to prevailing environments. Also known as heterozygous regulation.

Multifunctional Polymorphic Controller (MPC) - those parts of the genome which create additional capacity through interaction of polymorphisms controlling and regulating cascades and networks.

Major Histocompatibility Complex (MHC) - is the genomic region on 6p including the HLA loci.

Mutation - a change from the inherited form.

Paralogy - duplicated regions now on separate chromosomes.

Penetrance - the degree to which genes are expressed depending on sex, hormonal status, age, infection and stochastic or reactive meshing.

Polymorphic Frozen Blocks (PFB) - of genomic sequence rich in duplicons, RLEs, indels and polymorphisms and shuffled *en bloc* by meiotic recombination.

Polymorphism - inherited and conserved differences between individuals.

Quantum genomics - distributed packages of polymorphic sequence, including the basic agents of evolution such as duplicons, indels and Retro Like Elements.

Recessive - the expression dependent on both maternal and paternal alleles or haplotypes.

Retroviral Like Elements (RLEs) - including Alus, LINES and SINES, often referred to as transposons, retroelements, repeat or repetitive genomic elements.

Single Nucleotide Polymorphism (SNP) - a term used commercially to refer to single base difference whether inherited or recent.

Speciation - the process by which a demonstrably different population emerges and stabilizes.

trans - relating to the corresponding chromosome.

Wagyu - literally Japanese cow, includes Japanese Black, Red/Brown (Akaushi), polled, Mishima, Kobe beef. In general, the crosses between Mongolian and European cattle for two hundred years with later infusion.

Chapter 1

An Introduction to Evolution

Malthusian catastrophe or anamnestic evolution?

- *Quantal evolution.*
- *Ancestral haplotypes regulate functions and disease.*
- *The importance of duplication.*
- *Speciation and the importance of retroviral integration.*
- *The quantal concept of evolution.*
- *The practical implications of quantal genomics and evolution.*
- *Acknowledgements and references*

One of the great challenges in evolution is to explain how all 8 billion humans of today are different from each other. As demonstrated in subsequent chapters, there is incontrovertible evidence confirming the uniqueness of all these individuals.

It is often asserted that the differences are due to random mutations as a consequence of errors in copying the DNA sequence from parent to child. Hitherto there has been no other explanation. However, if mutations were the answer, there must be inexorable divergence and eventually some limiting threshold of incompatibility such as infertility and eventually extinction. This Malthusian view of evolution suits the purposes of doomsayers and proselytizers.

Another even greater challenge is to understand how, in spite of uniqueness, all these individuals retain essential similarities, which protect their ability to survive and procreate. Witness the success of the 8 billion individuals and, necessarily, all of their ancestors, over at least 10,000 generations.

Thus there is a paradox to be resolved. How can the apparent antitheses of individuality and compatibility co-exist and be guaranteed generation after generation? The answer is to partition the genome. Some parts create and protect uniqueness whilst others provide those structural elements that allow survival and reproductive success. Both parts need to be conserved, faithfully inherited and able to withstand any errors in copying, but especially, of course, the individuality component. Both parts provide a form of ancestral memory captured in the term “anamnestic evolution”.

As a further solution to the paradox, the individuality component, referred to here as the Multifunctional Polymorphic Controller (MPC), actually interacts and controls the remainder of the genome, providing additional options for survival.

Quantal Evolution

The term Quantal Evolution refers to protective packaging of special DNA sequences, which ensure that each individual is unique but, at the same time, equipped with the experience of ancestors.

There are hundreds of such packages, best known as Polymorphic Frozen Blocks (PFB), accounting for some 10 percent of the genome. Frozen refers to the fact that the sequences are protected from mutation and recombination.

In each individual there is a paternal and a maternal sequence at each block but, within the whole population, there are hundreds of alternative sequences. It is these protected and conserved Ancestral Haplotypes, not mutations, which create diversity.

Surprising to many:

- *Individuality is inherited.*
- *Polymorphism is conserved.*

Given hundreds of independent blocks and hundreds of options at each, all that is required is random mating and remixing by meiotic recombination which occurs between the blocks and therefore blends sequential Ancestral Haplotypes into new *cis* combinations whilst conserving the sequence of each. Thus, at each fertilization, there are sufficient combinations to ensure that every individual is unique. This process is described in more detail in subsequent chapters.

Ancestral Haplotypes Regulate Functions and Disease

The discovery of Ancestral Haplotypes was unexpected and largely incompatible with the thinking of the time. There appeared to be no way that very long sequences could be inherited *en bloc* from very remote ancestors? In any event, why would they?

The prevailing wisdom was based on Population Genetics, the Modern Synthesis, Neo-Darwinism and the infinitesimal model of free recombination. Thus, at meiosis, maternal and paternal genes were exchanged. Any departure from free recombination was designated "linkage disequilibrium" as might happen if, for example, selection of one gene resulted in hitch-hiking of those nearby.

In fact, as initially shown for the MHC, there were highly predictable haplotypes of 1 million base pairs containing hundreds of genes. These had to be Ancestral Haplotypes because they were the same in very distant relatives separated by thousands of generations. The implication was clear: These sequences had been protected from recombination and mutation.

On further investigation, we found that these Ancestral Haplotypes were defined by the non-coding regions rather than the genes they contained. At the time it was widely believed that non-coding sequences were "junk" or "selfish" and of no relevance to function let alone evolution. This extraordinary misconception was driven by adherence to the simplistic models of the Neo-Darwinists. Genes mutate and their mutant products are selected by natural selection. To the contrary, the noncoding regions are the key to gene regulation and therefore determine how and when genes are expressed.

Ancestral Haplotypes regulate function and disease. For example the 8.1 haplotype sets the expression level of inflammatory mediators, such as tumour necrosis factor, complement, etc., and determines the severity of numerous, apparently unrelated, diseases. It can only do this by having remote effects on numerous genes. That is to say, the polymorphisms that together define the AHs are actually recognized elsewhere in the genome.

There is a simple explanation for the importance of such sequences. Polymorphisms interact. Consider two steps in a pathway, such as a protein cascade. If each protein occurs in multiple forms, the consequences will be determined by the fit between the two. In this way differences between two subjects can be magnified. Greater options give a certain robustness and an ability to adapt to changing environments. Polymorphism and heterozygosity can be advantageous to the individual as well as the population. The Multifunctional Polymorphic Controller (MPC) is described in more detail below.

The Importance of Duplication

In exploring the structure and function of Polymorphic Frozen Blocks, there were many surprises. One of the biggest was the amount of duplication. Segments of sequence were imperfectly replicated often many times over. Of course it had been known for decades that gene duplication occurred and that increased copies could be beneficial. For example, in the Complement System, the number of C4 genes is the major determinant of how much C4 the individual produces. Further, if there are multiple copies, these can diverge at speciation so as to result in different alleles. There are many different forms of C4. Thus the idea of beneficial duplication became firmly embedded in Neo-Darwinian evolution. The surprises were that the same phenomenon had occurred in the non-coding regions of Ancestral Haplotypes leading us to speculate that extra copies of regulatory elements or perhaps any incidental sequence may also be functional and beneficial. More "junk" is better "junk"!

In fact, there are many levels of duplication. In addition to the imperfect segments found within Polymorphic Blocks, it became increasingly obvious that whole blocks can be duplicated locally or remotely. Commonly, this phenomenon of paralagous duplication is referred to as Whole Genome Duplication as appears to have happened in the evolutionary development of early vertebrates such as amphibia and teleosts. Paralogous regions can be tracked thereafter through mammals although the divergence increases and the events become somewhat blurred. However the changes are very largely species specific. It appears therefore that whatever events occurred in early vertebrates, paralogy is refashioned at each speciation event. As shown below, there are some uncanny similarities and yet some major differences when humans and cattle are compared. One conclusion is that paralogy provides extra options and therefore redundancy combined with opportunity.

Speciation and the Importance of Retroviral Integration

In the minds of many experts, there is the notion that speciation is related to physical barriers to fertilization. Here, however, we are more interested in creating and conserving change as patently occurred in defining humans hundreds of thousands of years ago. It appears very likely that this outcome was slow in the making and probably iterative. Eventually the ancestors of all humans emerged and then flourished.

The process was associated with radical changes in Polymorphic Frozen Blocks (PFB). These changes are in addition to the reorganization of chromosomes and redistribution of paralogous regions. They differ in one very important respect. Within PFB, hundreds of AHs were generated. The whole diversity of humans was very largely fixed from that time. So within these blocks there must have been extraordinary plasticity (or "polymutations") with subsequent differences in copy number, orientation, alignment, insertion, deletion and polymorphism - all of which are specific for a particular haplotype. How to achieve such a revolution of the previous status quo? This could not have occurred within a single ancestral population of tens or hundreds of thousands. Rather, there must have been successive expansions and contractions allowing the population to form from disparate forebears in order to achieve the diversity now observed.

One explanation is that there were successive retroviral infections which reduced the population size but also refashioned the PFB. This type of process could be possible if there were steps such as infection followed by vertical transmission of integrated retroviral sequence which created a degree of immunity to closely related viruses. Given that HIV is known to integrate into transcriptionally active gene regions, gene clusters resulting from retroviral-induced, duplicative expansions become focal targets for further retroviral integrations, an evolutionary process which will feed on itself. Thus small duplicated clusters can grow over time into longer haplotypes containing many imperfectly duplicated segments. This model is described further below.

There are several other interesting aspects to the evolution of PFBs. There could be a way of achieving great diversity in even a small population given the redundancy of gametogenesis and fertilization and the potential for selection at that stage.

However generated, there must be a process which eliminates further change. It is as if the PFB captures polymorphism and then, upon a certain limit, protects what has been achieved. Mutation must be suppressed. Intrablock recombination must be prevented. There are models which show that polymorphism can be self sustaining.

An interesting feature of PFBs is that they are scattered around the genome and separated by almost monomorphic sequence. This architecture permits protection of the content whilst allowing recombination between the blocks. In this way the influence of the MPC is distributed while the options for *cis* combinations are retained.

Here we focus on the formation and protection of individuality with its downstream benefits but there must have been other revolutions occurring at the same time. For example, the development of the human brain was contemporaneous and must have created requirements for yet more exquisite control of the pathways involved. Somehow, evolution managed to refine control whilst exploiting the advantages gained from previous species.

The Quantal Concept of Evolution

The major changes which resulted in humans were complex and undoubtedly risky. At each step, pre-existing retroviral immunity failed resulting in successive epidemics. Just as the retrovirus exploited weakness so did it accelerate genomic rearrangements and particularly duplication, insertion and deletion. Trial and error established an effective symbiosis between host and virus. At the same time unique Ancestral Haplotypes were established. These were selected for their ability to interact and control complex pathways and networks.

Eventually a small population emerged and succeeded in reproducing. Although the phenotype was largely maintained, subpopulations migrated away and adapted to new challenges including diverse climates. Over hundreds or thousands of generations, some sections of AHs were shuffled but the basic elements were retained. The revised AHs appear in greater or lesser frequency in the subpopulations or “ethnic groups” of today but can be tracked backwards. They are associated with phenotypic features which define recent ancestry largely due to the effects of the MPC in quantitative regulation of, for example, hair colour, height and endurance. Recent ancestry is a continuum in that it explains how unique individuals nevertheless retain degrees of family likeness, disease susceptibility and patterns of behaviour. Note that all this diversity exists without the need for ongoing mutation. Nor is there a role for ongoing selection. Indeed, quantal evolution explains why it has been so difficult to demonstrate examples of natural selection. The job was very largely completed by the time the first successful ancestors emerged.

Perhaps the most important concept underlying quantal evolution is that advantages are protected. In the case of PFB, long polymorphic sequences have multiple influences through gene regulation. If recombination and mutation did occur within the ancestral haplotype, there would be diverse consequences on many pathways and networks. At the stage of fertilization and early embryonic development, some form of purifying selection could result in losses. Thereafter, however, it is simply necessary to protect the intact haplotype from disruption. It will be interesting to determine precisely how this conservation is maintained but, in the meanwhile, quantal genomics contributes a series of testable hypotheses.

The notion of the MPC extends beyond quantitative regulation in a fixed environmental setting. Given interactions between multiple polymorphisms around the genome, it becomes clear that there is the potential for inherited anamnestic adaptation. Thus rather than imagining, as has been universal, that adapting required genetic change such as mutation, the capacity is conferred by virtue of meshing between conserved polymorphisms. Wherever interactions occur, heterozygosity permits options not available to homozygotes. The more interactions, the more the combinatorial differences by chance alone. If, in addition, the options are affected by environmental factors, there are real prospects for adaptation within the individual. For example, harsh conditions leading to starvation may favour some options and, by analogy, lead to a gear change (Figure 1).

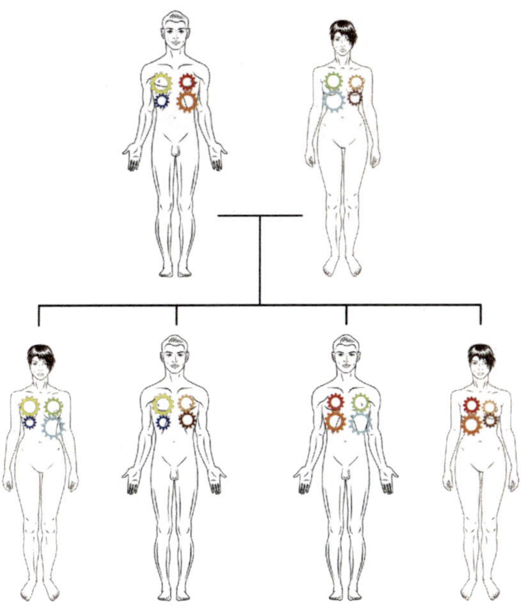

Figure 1
Segregating ancestral haplotypes interacting as meshing gears.

Ancestral haplotypes are represented by two cogs meshing vertically (in *cis*). For example the father (*ab*) has yellow-blue (*a*) and red-orange haplotypes (*b*). These have been transmitted to older daughter (*ac*) and the older son (*ad*).

The general view of today seems to be that humans have been selected for each environment as it changes over time. Take climate for example. If humans had been finely selected for today, how did they cope with the previous cycles and how could they survive future cycles. The MPC solves this dilemma. There is inherited robustness. There is anamnestic capacity. Similarly with respect to infections and somatic mutation leading to cancer.

In spite of some optimism arising from the above there will be a need to maintain some form of symbiosis with retroviruses in particular. Clearly there is a pressing need for improved understanding of this issue and hopefully of how to manage the risks. The best approach will be found by understanding the past. Some generalisations are now possible:

- Humans are not predestined for immediate disaster.
- Individuality is a necessary consequence of complex systems.
- Complex systems need not be perfect; imperfecttions in interactions can provide options.
- Most current teaching and practice of evolution is focused on the less informative part of the genome.

The Practical Implications of Quantal Genomics and Evolution

Monogenic traits and diseases

These can be regarded as the exception which proves the rule. Mutations which survive in spite of all protective and repair processes can be demonstrated in some families with this type of disease. The preconditions are that:

The disease is relatively rare so as not to have major effects on the larger population. The more common, the less important.

The mutation affects a gene which is functionally important but not critical to survival of immediate progenitors. Thus, for example, many behave as recessive disorders in that disease is only expressed in the homozygote.

Expression is often delayed and contingent on other factors. Penetrance and severity can vary.

The mutation occurs in regions where protection and repair are inadequate – generally, between rather than within, PFB. This issue is addressed in companion chapters.

Polygenic traits and diseases

These so-called "common" diseases are characterized by:

- Onset in later life
- Degrees of severity
- Sex related differences
- Complex genetics often with low penetrance
- Differences depending upon intermediate ancestry

All of these features are consistent with a role for the MPC. Obviously the interactions will be complex and difficult to identify using the concepts of population genetics. Indeed the disappointments to date could have been anticipated. The weaknesses of approaches based on SNPs and GWAS are exposed in later chapters.

Transplantation

The current convention emphasizes rejection due to immune reactions. Whilst these are important - especially if there is presensitisation - other issues should be considered. The fact that grafts do better when the donor and recipients are well matched may be partly due to the fact that they share similar MPCs and therefore coordinated gene regulation.

Livestock breeding

There has been great disappointment in the effectiveness of artificial selection based on the teaching of population genetics. A major problem is the dogma that there are critical genes freely recombining. Again SNPs and GWAS have been promoted for commercial reasons. A more pragmatic approach is to blend the relevant ancestral haplotypes.

Forensic investigation

As discussed below, there is far too much emphasis on overly simplified thinking and testing. The consequences can be serious.

Legal ownership

Whilst it is implicit in law that each individual is unique and responsible, disputes are yet to be based on direct and reliable evidence.

Individuality

The notion that individuals are created by errors seems to have encouraged many to treasure themselves and to disregard the degree to which they are simply remixes of the past. The "social sciences" tend to ignore the strength of anamnestic evolution, the inherited restraints on an individual and often exaggerate other possibilities, such as freewill and nurture, as if each individual were a *de novo* creation.

Terrorism

Identification of individuals can be vastly improved. Simple DNA testing can identify ancestral haplotypes.

A more complete history of the discoveries outlined here and their role in understanding genomics is provided by Dr.. E. J. Steele's recent review [1] and concurrent book entitled "Ancestral Haplotypes" [2].

References

1. Steele, E.J. 2014 Reflections on Ancestral Haplotypes - Medical Genomics, Evolution and Human Individuality. *Perspectives in Biology and Medicine*. Vol 57 (2). Spring 2014.pp 179-197.

2. Steele, E.J. 2015 *Ancestral Haplotypes: Our Genomes Have Been Shaped In The Deep Past*. Nearurban, Dallas, Texas. ISBN 978-0-9864115-0-2

****

Chapter 2

Medicine, Immunology and Transplantation

Benefits of Clinical Science in Demonstrating that the Major Histocompatibility Complex is a Multifunctional Polymorphic Controller and a Model for the Genome

- *Mutation to polymorphism to genomic regulation.*
- *The MHC and quantum genomics.*
- *Why was the MHC so useful in discovering the rules of genomics?*
- *Why were the discoveries so resisted by geneticists, molecular biologists and neo-Darwinists?*
- *Implications of HLA and MHC genomics for clinical medicine and biology.*
- *The MHC allows the demonstration of the Multifunctional Polymorphic Controller (MPC).*
- *Implications for individuality.*
- *Implications of HLA and MHC genomics for evolution.*
- *Where does genomic variation come from? Is it polymorphism or mutation or both?*
- *Acknowledgements and references*

Indeed in the context of "blending inheritance" Charles Darwin in 1857 said to Thomas Huxley "... *a sort of mixture....of innumerable individuality, as each parent has its parents and ancestors... I can understand no other....way .crossed forms go back to ancestral forms*" (www.darewinproject.ac.uk/entry-2166).

Mutation to Polymorphism to Genomic Regulation

The meticulous descriptive approach of the HLA workshops has been rewarded. Although often portrayed as clinical, technical and quaint, it has become clear that the discoveries arising from HLA serology actually anticipated those often attributed to molecular techniques and eventually to sequencing. Further, irrespective of technology, the generalizations which describe the MHC region also apply to the whole genome. Far from being the oddity, the MHC is the prototype.

Shown in Table 1 is an abbreviated summary of these lessons from MHC genomics.

Table 1 Lessons from MHC Genomics

- Human diversity is inherited from ancestors rather than created by recent mutation.
- Diversity is regenerated at speciation and maintained by meiotic recombination between the ancestral haplotypes of near-by polymorphic frozen blocks.
- The unit of inheritance is the ancestral haplotype.
- Such sequences carry specific alleles, duplicons, indels and RLEs which together regulate gene expression.

The MHC and Quantum Genomics

The MHC, as used here, refers to that region of 6p which contains diverse HLA class I sequences telomerically and HLA class II sequences centromerically (Figure 1).

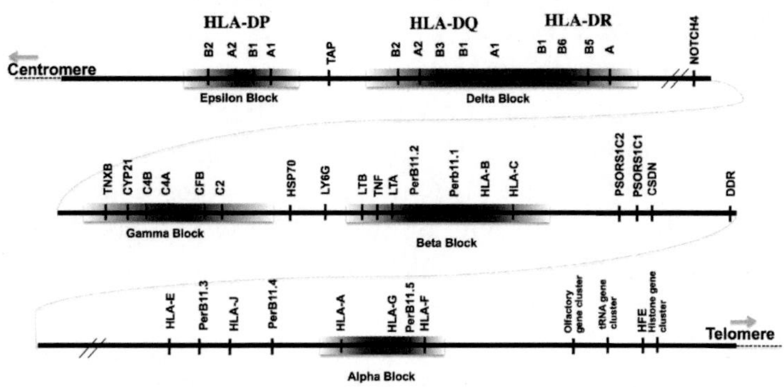

Figure 1
The organization of the MHC provides a model for the genome.
Each ancestral haplotype has its own map. Adapted from Ref. 20. PFBs are shaded. Not all genes are shown. PerB11 is now designated MIC.

Within these several megabases, there are blocks of intense polymorphism separated by sequences which are essentially invariant apart from some microsatellites. At each of these blocks, which we refer to as alpha, beta, etc., there are alternative sequences throughout. That is, there are extensive continuous haplotypes which have been inherited faithfully over many generations and are therefore ancestral. These polymorphic sequences are conserved irrespective of the apparent functional significance of the components. Thus, the conservation applies equally whether the sequence is classified as intergenic, coding, noncoding or regulatory and whether it contains indels, duplications or retroviral like elements.

The discovery of polymorphic frozen blocks had some important implications. Previously, sequence differences had been thought to occur randomly rather than in packages or quanta. Clearly, however, polymorphism is not fragmented by meiotic recombination; rather crossing-over occurs between blocks and permits the inheritance of intact haplotypes. These may be hundreds of kilobases in length thereby permitting inheritance of huge and complex sets of alleles at genes surrounded by regulatory and other modifying sequences including duplicons, indels and RLEs.

These features are now known to be characteristic of other genomic regions [1-3] and are being re-discovered daily although generally without knowledge of the first descriptions in the MHC more than 30 years ago. Terms such as "archaic haplotypes" and "islands of linkage disequilibrium" are now used by some and undoubtedly depend upon the same Polymorphic Frozen Blocks seen through different eyes, such as those of the SNP industry.

Why Was The MHC So Useful in Discovering The Rules Of Genomics?

We may choose to emphasize the quality of the initial serology and its meticulous analysis or the importance of an agreed unequivocal nomenclature. These were critically important but there are several additional explanations for the success of the HLA fraternity in describing genomics before the term existed.

The HLA fraternity functioned as a cohesive and productive group. Quality data were essential. Discoveries were quickly evaluated and confirmed or rejected. Cooperation was assumed and valued. Multicentre studies were the rule. These attitudes were developed and enforced by the pioneers. There has been nothing quite like it; before or since. By contrast with the HLA fraternity, some other areas of genetics were characterized by vitriolic "beanbag" debates, many of which continue today, fuelled by adherence to the concepts which were developed for flies and worms in the pre-genomic era.

Another important reason for the success was the immediate relevance to biological endpoints. Through transplants and disease association it became obvious that HLA typing had real significance. Unlike population genetics and evolution, there was no need for assumptions and modelling in order to determine whether the typing was important [4-6]. As stressed by one of the pioneers, Paul Terasaki, the individuals needed results because they shared the motivation of successful clinical outcomes in transplantation and ultimately organ sharing.

To achieve their purpose, the different laboratories needed a technological breakthrough to ensure quality data irrespective of the testing centre. Terasaki microdroplet typing trays were the vehicle which permitted pooling of results. The use of human sera with high affinity antibodies provided another advantage. The epitopes were those seen *in vivo* after immunization by pregnancy, for example, and were therefore of biological and evolutionary significance.

Fortunately, thanks largely to the Bodmers, it was appreciated from the outset that family studies were critical in defining paternal and maternal haplotypes. For this reason, the MHC has been blessed by observing inheritance directly, as is so apparent from the work of van Rood and the Europeans. This strategy has avoided the artifacts which arise by attempting to derive haplotypes without family data and is the reason that the HLA fraternity discovered that ancestral haplotypes contain indels, retroelements and duplicons. It is also the reason that we recognized polymorphic frozen blocks long before the whole genome was sequenced. Serologically defined haplotypes could be compared thereby revealing that all 8.1 ancestral haplotypes share specific genomic features within the blocks.

Ancestral haplotypes explain why extreme linkage disequilibrium (LD) is found between certain specific polymorphisms [7-9]. The alleles are components of a conserved sequence [10-15]. Common haplotypes with common alleles will travel through generations yielding high delta values between those alleles. However, LD calculations are unreliable when haplotypes have rare or poorly defined alleles; it is still necessary to determine phase through family studies which, as expected, have helped to improve the definitions of alleles themselves. Further, empirically it has been found that LD analysis is inappropriate when there are innumerable loci, as is the case with ancestral haplotypes of many kilobases. [16, 17].

If the techniques of the day, such as SNP typing, ignore alleles, indels, duplications and RLEs, the haplotypes are poorly defined and generally confused with each other, especially when the frequencies differ greatly. Without any doubt, LD between selected nucleotides is very inefficient in identifying ancestral haplotypes [17]. There are too many alternative polymorphisms of too many types at too many sites. Observation of haplotypes in ancestral groups is much more accurate and explains why the HLA fraternity solved structural genomics long before the SNP consortia.

As a consequence of family analysis of the MHC, homozygous typing cells were routine from the 1980s leading immediately to comparison of extensive genomic sequences. Starting in 1985 and largely completed by 1992, the HLA group effort had revealed most of what is now known of genomic structure generally [18-27]. It would have been just if the HLA pioneers had been there when Clinton and Blair revealed the human genome to the world!

Another key advantage of the HLA workshops was that there were excellent laboratories from all regions of the world. Ethnic comparisons were routine over 40 years and have been extended since. For this reason, HLA typing has provided maps of human migration, estimates of population sizes, explanations for clines of disease susceptibility and, importantly, effective donor banks for transplantation and bone marrow grafts. All of these on shoestring budgets!

Why Were The Discoveries So Resisted By Geneticists, Molecular Biologists And Neo-Darwinists?

In spite of the obvious advantages of the MHC as an entrée to meaningful genetics, the findings have been very largely rejected or ignored. Commonly, one heard that the MHC was an aberration or a curiosity of no general significance or, at best, a frustration to unnatural transplantation. MHC "tissue typing" was often considered an arcane form of occupational therapy for ageing serologists!

Table 2 Alternative Dogma

- Genetic diversity or inherited variation requires ongoing mutation.
- Diversity accumulates through errors in copying of DNA.
- The unit of inheritance is the allele.
- At each locus alleles may be deleterious, beneficial or neutral.
- Meiotic recombination scrambles maternal and paternal alleles.
- Linkage disequilibrium between SNPs defines haplotypes.

Perhaps, in retrospect, the discoveries and rules shown in Table 1 were not communicated to a sufficiently wide audience. Still today, many experts in genetics are more likely to subscribe to an alternative set of dogma as outlined in Table 2.

These widespread misconceptions can be traced to underlying concepts. Most seem to be based on the idea that there is one standard human genome with occasional errors leading to SNPs, which affect or mark coding changes in alleles. This model has become widespread, fanned by commercial promotion of turnkey packages which measure and analyze the SNPs.

With such a mindset it would be difficult, if not impossible, to appreciate the importance of MHC or HLA typing and it must be challenging to interpret genomic studies which demonstrate the presence of unique individual genomes carrying conserved, functional ancestral haplotypes. And, impossible for those who adhere to the concept of "junk".

Gillespie's text book on population genetics - defined there as "the genetic basis of evolution" - asserts that "insights are theoretical rather than observational …" and it is not important that .. "they could not be verified in a natural setting" [28]. This assertion contrasts with the meticulous descriptive approach of the tissue typing and HLA fraternity and explains the difficulty in penetrating population genetics and, indeed, the resistance to using the MHC to elucidate the genetic basis of evolution.

The main challenge for population genetics and neo-Darwinism is the finding that extreme HLA polymorphism is effectively blended *en bloc* rather than created by on-going mutation. Thus, the reality reverts to proto-Darwinism in that there is continuing mixing of the variants which survive speciation with or without modification. These polymorphisms define the new species and control its survival until the next speciation event.

Implications of HLA and MHC Genomics for Clinical Medicine and Biology

How do MHC genes influence diverse functions such as

complement activation and also determine susceptibility to hundreds of different diseases as different as Narcolepsy, Hemochromatosis, IgA deficiency and Insulin Dependent Diabetes Mellitus (IDDM)? In the case of myasthenia gravis the effect of the MHC is more relevant to the associated thymic hyperplasia rather than to the presence of autoimmunity itself. Even after 40 years of cataloguing such remarkable associations, there are very few simple answers [20]. However, there are some important generalizations as can be listed in Table 3.

These are important insights which might teach all of those attempting to understand complex traits, polygenic diseases, ageing, gene regulation and epigenetics, methylation, copy number variations and the influence of sex hormones. They certainly warn against applying the tools of population genetics and of the Neo-Darwinists. They also indicate that there is a huge gap between understanding structure, as we hope to soon, and function, as we should when the interactions between the structural elements are defined.

The MHC Allows the Demonstration of the Multifunctional Polymorphic Controller

The fact that most diseases are associated with ancestral haplotypes rather than one particular HLA allele was clear 30 years ago [7, 9, 29]. This implied that the explanation involves quantitative regulation of multiple functions. One of many examples is given by myasthenia gravis. The 8.1 ancestral haplotype is associated with thymic hyperplasia in females rather than the autoimmune disease itself.

The concept of a multifunctional polymorphic controller emerged in an attempt to explain how so many functions and diseases are regulated in a quantitative fashion by ancestral haplotypes containing predominantly

Table 3 Lessons from MHC

- Age and sex related penetrance implying gene regulation of multiple pathways including many of no relevance to immunology.
- Susceptibility to disease is carried by extensive ancestral haplotypes; a few monogenic diseases are exceptions.
- Ancestral haplotypes determine clinical subsets, associated features and rates of progression.
- Important candidate sequences include indels, retro-elements, and variation in copy number rather than polymorphism within coding regions.
- Genetic redundancy is characteristic and permits advantages.
- Unlinked paralogous regions may contribute to complex functions and survival.
- Polymorphic control of expression, penetrance and regulation will be important clinically and biologically.

non-coding polymorphisms largely due to imperfect duplication, retroviral-like elements and insertion and deletion. We now know that such genomic features are found in other PFBs outside the MHC. How can these regulate different pathways? Is the multifunctional polymorphic controller distributed around the genome? What is the evolutionary value of inherited polymorphism?

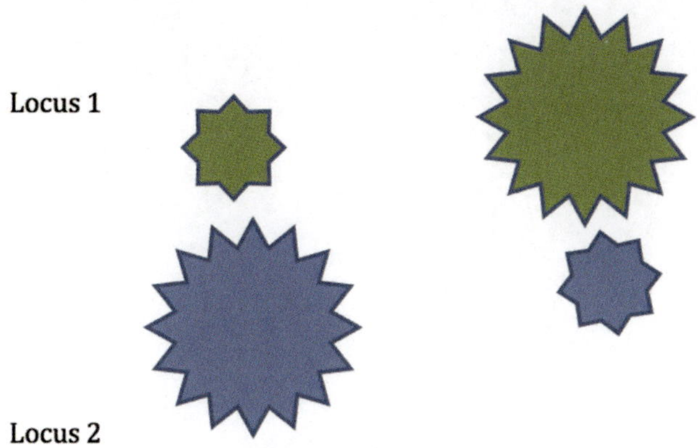

Figure 2
Interactions in *cis* and *trans* represented by meshing gear cogs.
Colour represents locus, size and shape represent alleles. See also Figure 3 below and Figures 4,5 in Chapter 4.

As described further below and elsewhere, the concept is disarmingly simple. Polymorphisms at one site react with polymorphisms elsewhere creating combinatorial options (Figures 2, 3 and Chapter 4). We expect that the benefits involve *stochastic blending* or *meshing* of complex, largely *paralogous cascades* which can then be expressed in degrees within a substantial range implying *reserve capacity* and therefore an *inherited potential to adapt.* By *stochastic blending*, we refer to polymorphic messages or products which are sufficiently similar to allow promiscuous interaction but different enough to create degrees of outcomes. By *paralogous cascades* we refer to messages or products of genomic regions which arose by segmental duplication but are sufficiently divergent to allow complex cascades such as, for example, the complement pathway. Such cascades bring efficiency of coordinated responses but also introduce the risk of impairment in the form of disease.

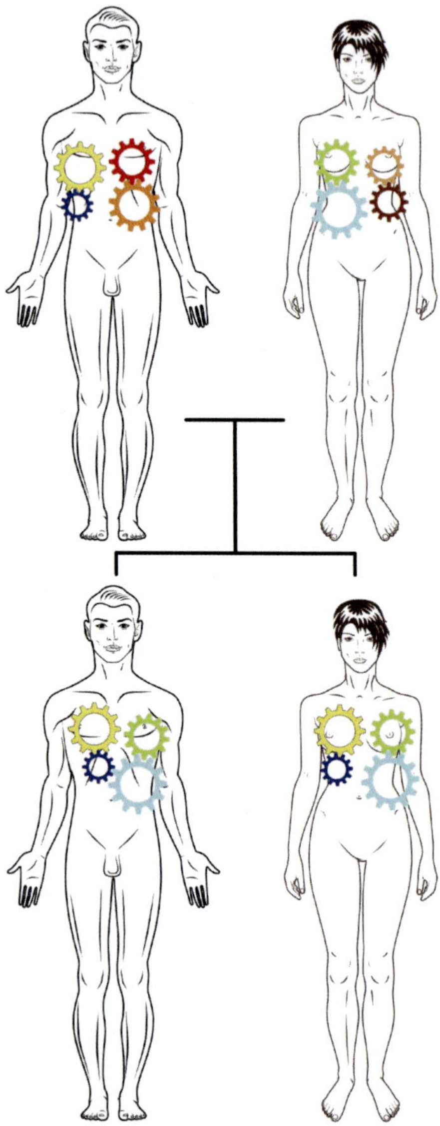

Figure 3
Reactive meshing dependent on hormonal and other environmental influences.

Coloured cogs represent ancestral haplotypes. The size and shape of cogs represent polymorphism or variant forms of the blocks. The two offspring are genotypically identical but the interactions are different as a consequence of the sexual environment. See also Figures 4, 5 in Chapter 4.

Reserve capacity is achieved by redundancy largely due to segmental duplication and therefore associated with polymorphism. The consequence of multifunctional polymorphic control is an *inherited potential to adapt* to circumstances and therefore evolutionary advantage in that effective function is permitted *in spite of changes in age, disease and climate.* Importantly, the genomic structure provides a form of memory in the sense that the conserved sequences have interacted successfully through the challenges which confronted ancestors. Mutation is clearly not required for survival; to the contrary, the power of the polymorphic controller provides a degree of protection from any form of mutation whether germ line or somatic.

Increasingly, since first demonstrated for TNF and IgA more than 20 years ago [30-32], there have been reports of examples of the MHC controlling and regulating expression and function of genes outside as well as within the MHC [33-37]. In some cases, inexplicably, SNPs have been implicated without considering how it is possible to isolate the effect of a nucleotide from the entire sequence of an ancestral haplotype. Nevertheless, there are now so many examples that it is no longer surprising that ancestral haplotypes have effects on pathways encoded outside the MHC.

Mapping these regulatory genes to a particular PFB became possible once ancestral haplotypes were defined. For example, the early findings on myasthenia gravis [38] and insulin dependent diabetes [39] have been confirmed repeatedly [40].

Implications for Individuality

With the exception of monozygotic twins, each of the 8 billion alive, and each of their ancestors, are generally considered to be unique. Could this be true genetically? Most believe that random mutation creates individuality. However, in searching for unrelated but compatible bone marrow donors, it becomes obvious that individuality is inherited and, in this regard, encoded within only a few MHC PFBs together summing to only one megabase of sequence after excluding the intervening invariant sequence. How can this be? Could there be more than sufficient alternative sequences in just one megabase?

Firstly, at each block there are thousands of alternative haplotypes as judged by the combinations of alleles at multiple loci in *cis*.

Secondly, at the sequence level there are even more because of differences in the indels, duplicons and retro elements. These give rise to multiple shifts in alignments for transcription. Together these genomic differences create vast options for polymorphism.

Thirdly, the functionally relevant sequence is the diploid combination in *trans* rather than the individual haplotype.

Fourthly, the maternal and paternal haplotypes at each block are conserved and remixed at each generation because meiotic recombination occurs between the blocks. This is a remarkably efficient way of generating continuing but pretested and useful quanta of diversity and is, one imagines, one of the keys to evolution.

Fifthly, the ancestral haplotypes are frozen in that they are largely immune from recombination and mutation whilst the polymorphisms are immune from repair. This has been demonstrated by sequencing examples from disparate individuals who could only have inherited the haplotype from a very remote ancestor [23, 26, 41].

In these ways it is possible to explain conservation of polymorphism as the basis for the inheritance of individuality combined with family likeness. Diversity is created by remixing a vast number of alternative haplotypes rather than by changing the haplotypes themselves.

****

Implications of HLA and MHC Genomics for Evolution

Within the MHC fraternity, there have been arguments for or against the various particular models of evolution but we have often been hampered by the disconnect between the observed (as summarized here) and the expected based on the teaching of the supposed experts in evolutionary theory and population genetics. Reconciliations have been attempted but universally accepted only by ignoring the unexpected and unpalatable lessons from the MHC.

To the contrary, it is now obvious that the MHC is the window to genomics and evolution. In fact, as many groups have been showing over the last decade, there are numerous polymorphic blocks distributed or seemingly scattered around the genome and these follow the same rules long since established by MHC studies [1-3, 42-44]. They account for most of the polymorphism within the entire genome. The interblock regions are much less polymorphic. The implications have been lost or overlooked. The MHC and similar blocks hold the keys to understanding difference, individuality and evolution.

Where Does Genomic Variation Come From? Is it Polymorphism or Mutation or Both?

For the present purpose, polymorphism refers to inherited differences in sequence when two or more subjects are compared [45]. Inheritance may be inferred when complex polymorphic sequences are shared by remote relatives but should be confirmed in 3-generation family studies capable of revealing meiotic recombination. Some earlier definitions of polymorphism stipulated an arbitrary minimum cut-off frequency such as 1% or even 5% but this has proven inappropriate. Rare, inherited polymorphisms may be biologically important in combination if not individually.

Mutation refers to a *demonstrable* change in sequence as can be deduced or, in bacteria, observed directly [46, 47]. In vertebrates, mutation has been inferred, as for example in rare monogenic traits or disorders which seem to have appeared in members of one family within recorded times. It has also been proposed when change is merely plausible. If dated and localized to a site, "recent" or "point" may be used but generally loosely. Some use "random" to emphasize that the mutation is not evolution by divine design or creation.

A distinction between polymorphism and mutation is a practical necessity when considering MHC sequences. It is those which are inherited, not mutated, which are important biologically. Unfortunately, however, yet other supposed experts in evolution use mutation in a different sense and refer to any difference at any time and may include such deduced and diverse events as recombination, inversion or duplication, irrespective of when the event occurred and whether it has become heritable. For the purpose of clarity, these variations should be specified as separate processes.

One term which encompasses all differences might reasonably be "variation" as used by Darwin when he asked: "where does the variation come from?" However, "variation" has also been used to refer to phenotype and has been confused with "variable" in the sense of non-heritable or non-reproducible. Only reproducible and heritable variations influence evolution.

Confusion now abounds, even within the MHC fraternity, exacerbated by the widespread use of SNPs to describe a single difference even when, more correctly, it should be recognized as just one component of an extensive haplotype which segregates in family studies. Sadly, in some Genome Wide Association Studies (GWAS), it is assumed that every SNP is actually a mutation. Indeed, in population genetics it is frequent to find polymorphism and mutation used interchangeably. This unfortunate habit ignores the teaching of the MHC: Polymorphisms may be conserved over many generations and even successive species.

One way of reducing confusion is to accept that a mutation may be repaired or lost or modified but it may also persist to become a polymorphism. Polymorphic frozen blocks are rich in diverse forms of variation such as indels, RLEs and duplicons and all are conserved. It is as though the PFBs actually act as safe houses for all forms of variation including those captured by rearranging blocks.

Many different forms of polymorphism appear to have arisen by insertion-deletion and/or imperfect duplication and then become conserved or "frozen" after speciation [20]. Patently, the sequence differences are not repaired. Nor are further changes permitted. Mechanistic explanations for such freezing would have important implications for evolution, disease and individuality. Perhaps a particular degree of sequence sharing in *cis* or *trans* serves to inhibit repair, recombination and, indeed, any form of mutation.

By this reasoning, those mutations which occur between blocks are repaired, deleted or re-mutated and, in terms of their impact on evolution, effectively ignored.

An entrenched hangover from bacterial and fruit fly genetics is the notion that there is a particular reference or "wild type" and that, from time to time, faulty copying leads to mutants of that type. This concept was discarded by the students of the MHC who recognized that there were multiple alleles and haplotypes all of which were equally "wild" and equally, non-mutant. There is no single human genome to select as the prototype, in spite of some choosing to use Watson, Venter and Hosa NCBI 36. Polymorphisms have to be recognized by comparison with each other rather than a "wild type." Nevertheless, various groups continue to use their habitual terminology even when entirely incompatible with an understanding of the MHC.

The persistence of semantic confusion has inhibited the dissemination of the insights arising from MHC genomics and has allowed some to cling to some earlier interpretations of evolution. There are repeated assertions, as for example by Olson that "most variation in genomes is due to recent mutations" [48]. This is simply incorrect. In humans, at least, most variation is due to inheritance of stable haplotypes with accumulation of nucleotide differences, indels etc. Recent mutations are the exception rather than the rule. How else could haplotypes be maintained with minimal change for thousands of generations? By our reckoning many at least date to the Out-of-Africa emigration estimated by Stephen Oppenheimer [49] to be ≥ 80,000 years ago.

Recently, there has been an acceptance of the concept of "indel induced mutation" [50-52]. This traces back to our early studies cited above and summarized by Dawkins et al. [20] and more recently by Longman et al [22] as Quantum Genomics. Our observations showed that indels and sequence polymorphism - not recent mutations - are closely associated, arose during or survived speciation and have been conserved since. The same is true of "repeat sequences" or RLEs. The term "mutation" could give the impression that changes are ongoing.

Another consequence of semantic confusion is the assertion that evolution must be based on selection of errors of copying. Whilst this concept may be relevant to bacteria and invertebrates, it has no place in vertebrate evolution. It is inconceivable that the same errors are selected at each generation with such efficiency that haplotypes are conserved to the extent observed in the MHC. Rather, inherited polymorphism provides reserve capacity via the multifunctional polymorphic controller.

Acknowledgements

Many colleagues and particularly J.F. Williamson played a critical role in preparation.

We are so grateful to have been a part of the MHC fraternity and acknowledge the benefit of clinical science when compared to the more industrial approach taken by the SNP group. Recognizing that there is more genome to address and evolution to explain, we look forward to a great deal more interaction with other fields of genetics.

I acknowledge the major conceptual contributions of Chester Alper and Edmond Yunis [8, 17, 33], Walter Bodmer [4], John Trowsdale [37], Miro Radman [46, 47], Jean Dausset [53-55], Malcolm Simons, Jan Klein [14], Pierre Pontarotti [18], Dan Geraghty [23], Stephen Oppenheimer [49], Frank Ryan [56], Jim Schapiro [57], Calegero Caruso (see Chapter 4), and many former colleagues cited here such as Yurek Kulski [20, 27], Silvana Gaudieri [41] and Craig McLure [1-3]. It is worth reminding ourselves that the breadth of MHC effects were already hotly debated by Jon van Rood, Rene de Vries, and others more than thirty years ago [7, 9]. Special thanks to Peter Kay, Jim McClusky, Marius Giphart, Ekkehard Albert, Guy Grimsley, Jim Peter, Louise Smyth, Paul Cameron, John Armstrong, Peter Hollingsworth, Martyn French, Frank Christiansen, Trevor Cobain, Wenjie Zhang, Katsushi Tokunaga, Chanvit Leelayuwat, Lawrie Abraham, and Mariapia Degli-Esposti, all of whom contributed so much in the early battles.

References

1. McLure, C.A., et al., Genomic evolution and polymorphism: Segmental duplications and haplotypes at 108 regions on 21 chromosomes. Genomics 2013 102(1): p. 15-26.

2. McLure, C.A., et al., Extensive genomic and functional polymorphism of the complement control proteins. Immunogenetics 2005 57(11): p. 805-15.

3. McLure, C.A., et al., Indels and imperfect duplication have driven the evolution of human Complement Receptor 1 (CR1) and CR1-like from their precursor CR1 alpha: importance of functional sets. Hum. Immunol. 2005 66(3): p. 258-73.

4. Bodmer, J., et al., Leucocyte antigens in man: a comparison of lymphocytotoxic and agglutination assays for their detection. Nature 1966 210(5031): p. 28-31.

5. Terasaki, P.I. and McClelland, J.D. Microdroplet assay of human serum cytotoxins. Nature 1964 204: p. 998.

6. Van Rood, J.J., Eernisse, J.G. and van Leeuwen, A. Leucocyte antibodies in sera from pregnant women. Nature 1958 181(4625): p. 1735-6.

7. Dawkins, R.L., Christiansen, F.T., and Zilko, P.J. (Eds) Immunogenetics in *Rheumatology: Musculoskeletal Disease and D-penicillamine, including reports from the D-Pen - HLA 82 Workshop*. 1982. Perth, Western Australia: Excerpta Medica, Amsterdam.

8. Alper, C.A., et al. Extended major histocompatibility complex haplotypes in man: role of alleles analogous to murine t mutants. Clin. Immunol. Immunopathol. 1982 24(2): p. 276-85.

9. Dawkins, R.L., et al. Disease associations with complotypes, supratypes and haplotypes. Immunol. Rev. 1983 70: p. 1-22.

10. Tokunaga, K., et al. Extensive deletions and insertions in different MHC supratypes detected by pulsed field gel electrophoresis. J. Exp. Med. 1988 168(3): p. 933-940.

11. Dawkins, R.L., et al. Conservation versus polymorphism of the MHC in relation to transplantation, immune response and autoimmune disease, in *Molecular Evolution of the Major Histocompatibility Complex*, J. Klein and D. Klein, Editors. 1991, Springer-Verlag: Heidelberg.

12. Degli-Esposti, M.A., et al. Ancestral haplotypes are conserved population haplotypes with a unique MHC structure. In *HLA 1991*. 1992. Yokohama, Japan: Oxford University Press.

13. Kay, P.H., et al. Coexistence of an MHC chromosomal segment marked by HLA B17, BfS, C4A6, B1, DR7, and DQw9 in different ethnic groups. Hum. Immunol. 1988 23(1): p. 27-35.

14. Klein, J. Origin of major histocompatibility complex polymorphism: the trans-species hypothesis. Hum. Immunol. 1987 19(3): p. 155-62.

15. Zhang, W.J., et al. Differences in gene copy number carried by different MHC ancestral haplotypes. Quantitation after physical separation of haplotypes by pulsed field gel electrophoresis. J. Exp. Med. 1990 171(6): p. 2101-2114.

16. Ahmad, T., et al. Haplotype-specific linkage disequilibrium patterns define the genetic topography of the human MHC. Hum. Mol. Genet. 2003 12(6): p. 647-656.

17. Alper, C.A., et al., The haplotype structure of the human major histocompatibility complex. Hum Immunol, 2006 67(1-2): p. 73-84.

18. Abi Rached, L., M.F. McDermott, and P. Pontarotti, The MHC Big Bang. Immunol. Rev. 1999 167: p. 33-44.

19. Abraham, L.J., et al. Sequence differences between HLA-B and TNF distinguish different MHC ancestral haplotypes. Tissue Antigens 1992 39(3): p. 117-121.

20. Dawkins, R.L, et al. Genomics of the major histocompatibility complex: haplotypes, duplication, retroviruses and disease. Immunol. Rev. 1999 167: p. 275-304.

21. Leelayuwat, C., M. Pinelli, and R.L. Dawkins, Clustering of diverse replicated sequences in the MHC. Evidence for en bloc duplication. J. Immunol. 1995 155(2): p. 692-8.

22. Longman-Jacobsen, N., et al. In polymorphic genomic regions indels cluster with nucleotide polymorphism: Quantum Genomics. Gene 2003 312: p. 257-61.

23. Smith, W.P., et al. Toward understanding MHC disease associations: partial resequencing of 46 distinct HLA haplotypes. Genomics 2006 87(5): p. 561-71.

24. Stewart, C.A., et al., Complete MHC haplotype sequencing for common disease gene mapping. Genome Res. 2004 14(6): p. 1176-87.

25. Zhang, W.J., et al. Organization and evolution of C4 and CYP21 genes in primates: importance of genomic segments. Immunogenetics 1993 37(3): p. 170-6.

26. Horton, R., et al. Gene map of the extended human MHC. Nat. Rev. Genet. 2004 5(12): p. 889-99.

27. Shiina, T., Inoko, H and Kulski, J.K. An update of the HLA genomic region, locus information and disease associations: 2004. Tissue Antigens 2004 64(6): p. 631-49.

28. Gillespie, J.H. *Population genetics : a concise guide.* 2nd ed. 2004, Baltimore, Md.: Johns Hopkins University Press. xiv, 214 p.

29. Kelly, H., et al. Susceptibility to IDDM is marked by MHC supratypes rather than individual alleles. Immunogenetics 1985 22(6): p. 643-651.

30. Abraham, L.J., French, M.A. and Dawkins, R.L. Polymorphic MHC ancestral haplotypes affect the activity of tumour necrosis factor-alpha. Clin. Exp. Immunol. 1993 92(1): p. 14 - 18.

31. Christiansen, F.T., et al. Human natural killer (NK) alloreactivity and its association with the major histocompatibility complex: ancestral haplotypes encode particular NK-defined haplotypes. J. Exp. Med. 1993 178 (3): p. 1033-9.

32. Dawkins, R.L., et al. Some disease-associated ancestral haplotypes carry a polymorphism of TNF. Hum. Immunol. 1989 26(2): p. 91-7.

33. Alper, C.A., et al. Prospective analysis suggests susceptibility genes for deficiency of IgA and several other immunoglobulins on the [HLA-B8, SC01, DR3] conserved extended haplotype. Tissue Antigens 2000 56: p. 207-216.

34. Cobain, T.J., et al., The coexistence of IgA deficiency and 21-hydroxylase deficiency marked by specific MHC supratypes. Annals of the New York Academy of Sciences 1985 458: p. 76-84.

35. French, M.A. and Dawkins, R.L. Central MHC genes, IgA deficiency and autoimmune disease. Immunology Today 1990 11(8): p. 271-4.

36. Fehrmann, R.S., et al. Trans-eQTLs reveal that independent genetic variants associated with a complex phenotype converge on intermediate genes, with a major role for the HLA. PLoS Genet. 2011 7(8): p. e1002197.

37. Trowsdale, J. and Knight, J.C. Major histocompatibility complex genomics and human disease. Ann. Rev. Genomics Hum. Genet. 2013 14: p. 301-23.

38. Degli-Esposti, M.A., et al. An approach to the localization of the susceptibility genes for generalized myasthenia gravis by mapping recombinant ancestral haplotypes. Immunogenetics 1992 35(6): p. 355-364.

39. Degli-Esposti, M.A., et al. Ancestral haplotypes reveal the role of the central MHC in the immunogenetics of IDDM. Immunogenetics 1992 36(6): p. 345-56.

40. Vandiedonck, C., et al. Pleiotropic effects of the 8.1 HLA haplotype in patients with autoimmune myasthenia gravis and thymus hyperplasia. Proc. Natl. Acad. Sci. U S A 2004 101(43): p. 15464-9.

41. Gaudieri, S., et al. Sequence analysis of the MHC class I region reveals the basis of the genomic matching technique. Hum. Immunol. 2001 62(3): p. 279-85.

42. Williamson, J.F., et al. Novel sequence elements define ancestral haplotypes of the region encompassing complement factor H. Hum. Immunol. 2008 69(3): p. 207-19.

43. Bailey, J.A., et al. Recent segmental duplications in the human genome. Science 2002 297(5583): p. 1003-7.

44. Daly, M.J., et al., High-resolution haplotype structure in the human genome. Nat. Genet. 2001. 29(2): p. 229-32.

45. Dawkins, R.L., et al., Mutation versus polymorphism in evolution. Genomics 2013 101(4): p. 211-2.

46. Radman, M., SOS repair hypothesis: phenomenology of an inducible DNA repair which is accompanied by mutagenesis. Basic Life Sci. 1975 5A: p. 355-67.

47. Radman, M. SOS replication: a distinct DNA replication mechanism which is induced by DNA-damaging treatments? 1970. DNA Repair (Amst) 2005 4(6): p. 732-8.

48. Olson, M.V. Human genetic individuality. Annu. Rev. Genomics Hum. Genet. 2012 13: p. 1-27.

49. Oppenheimer, S. *Out of Eden* Constable and Robinson. London 2004.

50. De, S. and Babu, M.M. A time-invariant principle of genome evolution. Proc. Natl. Acad. Sci. U S A 2010 107 (29): p. 13004-9.

51. McDonald, M.J., et al. Clusters of nucleotide substitutions and insertion/deletion mutations are associated with repeat sequences. PLoS Biol, 2011 9(6): e1000622.

52. Tian, D., et al. Single-nucleotide mutation rate increases close to insertions/deletions in eukaryotes. Nature 2008 455(7209): p. 105-8.

53. Dausset, J. Iso-leuco-anticorps. Acta Haematol. 1958 20: p. 156-166.

54. Dausset, J., et al. Centre d'etude du polymorphisme humain (CEPH): collaborative genetic mapping of the human genome. Genomics 1990 6(3): p. 575-577.

55. Cann, H.M., et al. A human genome diversity cell line panel. Science 2002 296(5566): p. 261-2.

56. Ryan, F. *Virolution : the most important evolutionary book since Dawkins' Selfish Gene.* Harper Collins 2009 356 pages.

57. Shapiro, J.A. *Evolution: A View from the 21st Century.* 1st ed. 2011 Upper Saddle River, New Jersey 07458: FT Press 254 pages.

Chapter 3

Artificial Selection of Livestock

Benefits of the Quantal Model of the Genome to the Selective Breeding of Livestock

- *Blending two contrasting views of inheritance.*
- *Predicament facing breeders.*
- *Potential of comparative genomics.*
- *The genomic approach.*
- *Selective breeding of livestock .*
- *Domestication and development of breeds.*
- *Simple traits are not so simple.*
- *Complex traits are very complex.*
- *The example of Wagyu.*
- *Population or quantitative genetics.*
- *Marker assisted selection.*
- *EBVs.*
- *Heritability.*
- *Blindspots in applications of evolution to vertebrates.*
- *Reproductive biology.*
- *Genomics.*
- *An alternative strategy for selective breeding.*
- *Acknowledgements and references.*

Selective breeding of livestock had already achieved scientific respectability and commercial success a century before publication of the ORIGIN OF SPECIES and nearly two centuries before the availability of profession-

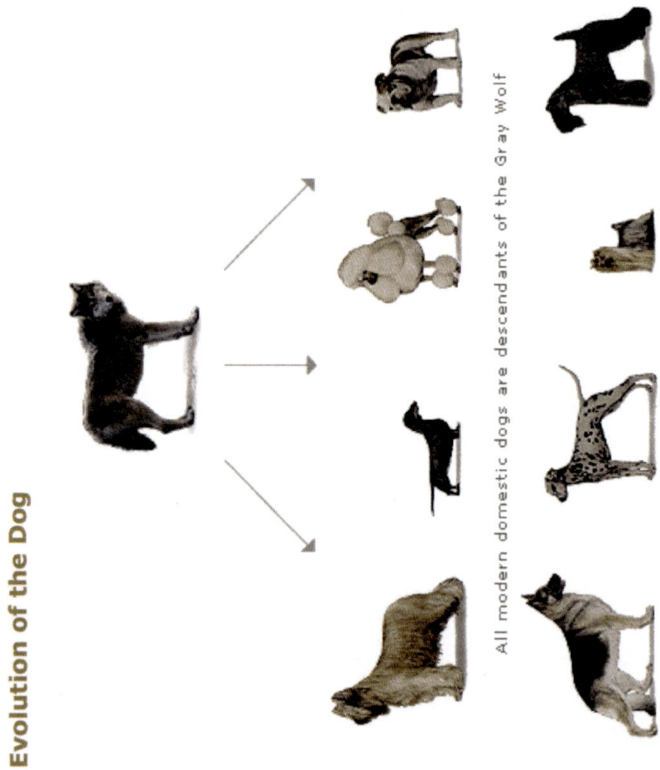
Evolution of the Dog
All modern domestic dogs are descendants of the Gray Wolf

al advice based on the "modern synthesis" of evolution and the development of quantitative genetics. Needing only careful observation and a few pragmatic principles, breeders continue to adapt diverse species including birds, fish and plants in addition to livestock. Advances in understanding evolution, heritability and genetics have come and gone without influencing or explaining the manifest success of artificial selection. There remains an intriguing contrast between the extreme diversity following artificial selection (see Figure 1) and the relative resistance to change in natural populations. In fact, breeders achieved not only rapid change but also striking stability (or conservation) thereafter. Breeds or "lines" were "consolidated" so that the advantages were retained. These achievements are all the more remarkable because they were based on accumulated experience and, even to this day, remain to be explained.

During the last several decades, breeders have been encouraged to modernize by purchasing "superior genetics" and adopting breeding plans set down by the experts in quantitative and economic genetics. See, for example Lush [1] and Falconer [2].

Why, then, is there so much skepticism amongst current livestock breeders? Why have dairy farmers increased milk production at the expense of health and fertility? Why did beef breeders have to discover that the commercial tests to assist in sire selection were prematurely promoted and actually misleading? Why have fine wool breeders withdrawn support for quantitative genetics? Why is there such public hostility to the pig and poultry industries? Are there solutions to these problems? If evolution and natural selection were truly understood, it would be reasonable to hope that there would eventually be some successful problem solving for the

Figure 1 Evolution of the Dog. All modern domestic dogs are descendants of the Grey Wolf. Dog Illustration by Chet Jezierski. Reproduced with the kind permission of the American Kennel Club (www.akc.org).

livestock industries. Has artificial selection been compromised by indiscriminate and erroneous adoption of some of the teachings of evolutionary and population genetics? Is the concept of "superior genetics" based on economics rather than genetics?

An authoritative review by Hill [3] addresses some of these questions and provides a much needed assessment of the assumptions implicit in advice offered to breeders : "When we look at ….. the application of population genetics to animal breeding we find that much of it has explicitly or implicitly been based on the infinitesimal model of Fisher 1918" [4] which "presupposes an effectively infinite number of unlinked loci" which associate transiently because of "free recombination".

Such assumptions are fundamentally flawed, as summarized below, but as Hill notes: "Notwithstanding the strong assumptions…. the Best Linear Unbiased Predictor (BLUP) of random effects… has become all pervasive in livestock improvement".

As Hill suggests, it appears that there is now some recognition of the flaws in the thinking which underlies BLUP and its derivatives like genomic BLUP [5] which, as Hill comments, "is equivalent to assuming that the distribution of marker-associated effects on the trait is the same for all markers fitted across the genome and each is sampled from a normal distribution with the same variance." However, whatever the emerging reservations of Goddard [6], Meuwissen and Goddard [7], Wray et al. [8] and Gianola [9], as cited by Hill, the most fundamental errors are yet to be addressed. These include a more general failure of population geneticists to address the lessons arising from genomics, especially with respect to block structure and all the associated implications. For example, in reality:

- The relevant genes occur in specific clusters, within polymorphic blocks.

• Ancestral haplotypes are conserved and inherited *en bloc.*

• Recombination is not random or free.

• Spontaneous mutation is suppressed within polymorphic blocks.

Thus, inheritance depends upon whether or not the descendant receives relevant, conserved haplotypes. By contrast, as foreshadowed by the quotes above, BLUP adopts the infinitesimal or 'lottery' model, assumes that both parents contribute equally at each generation and applies statistical methods. This distinction is fundamental but also practical in that, to be effective, the breeder must consolidate and conserve any advantages at each generation. To do so, it is necessary to have or develop haplospecific markers which allow inheritance to be tracked and understood.

Blending Two Contrasting Views Of Inheritance

The following figures (Figure 2 to Figure 6) are intended to be read seriatim.

Figure 2
Importance of clustering functional genes
Colours represent loci and numbers represent alleles at those loci. On the left is the basis of the 'Infinitesimal Model' used in Population Genetics. Free recombination between loci. On the right, loci are clustered within polymorphic frozen blocks (shown by alignment of loci). These are the important genes in terms of inherited differences because of *cis* and *trans* interactions. Single loci display simple Mendelian inheritance.

In Figure 2 the subject on the left is constructed according to the basic assumptions adopted by quantitative geneticists since 1918 [3]. Genes are shown as independent. The subject on the right gives a post genomics version with the same genes but some of these are clustered together into blocks and interact with each other to create greater complexity.

The next figure makes the critical point that a comparison of multiple individuals reveals extensive polymorphism within the blocks. Elsewhere in the genome, the situation accords with the teachings of population genetics.

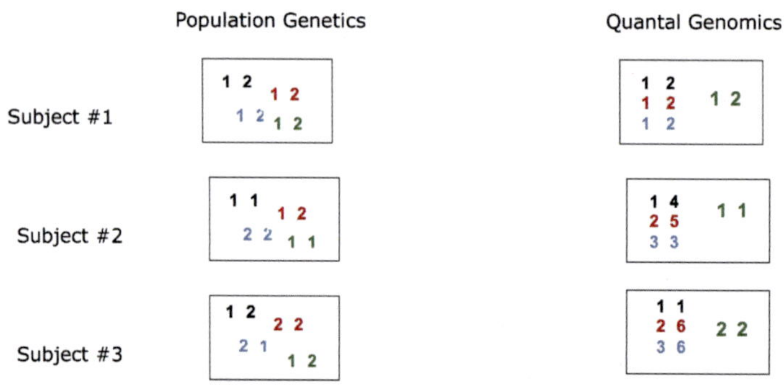

Figure 3
Clustering of polymorphism
On the left, three different individuals further showing the basis of the Infinitesimal Model used in Population Genetics. Note free recombination between loci showing homozygous and heterozygous loci in the same individuals. On the right, polymorphism is greater and clustered. Note, additional alleles. On the right, the green locus is outside the polymorphic frozen block.

Figure 4 shows the consequences of a mating of first and second genotypes from Figure 3. On the left, segregation is random and unpredictable. Inheritance is a lottery depending upon the interactions and regulation of the alleles actually received. Note also that although the offspring inherit half their alleles from each parent, the differences between the genotypes of the children and their parents can be much less than 50%. On the right, there are advantages in terms of diversity and predictability. Because the haplotypes are inherited *en bloc*, the number of available combinations is reduced. However packaging the major differences as haplotypes generates diversity between each offspring, and between offspring and their parents. Further, the haplotypes contain the non-coding elements required for interaction and regulation and, as a direct consequence, have the potential to determine phenotypes. The diversity is packaged, conserved, regulated and pre-tested or, to coin a phrase, selected *en bloc.*

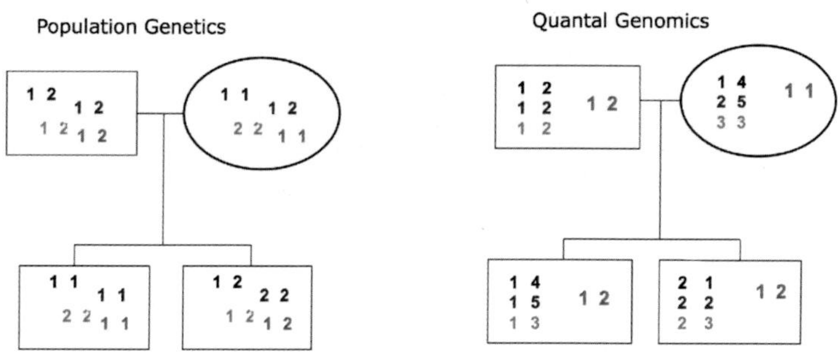

Figure 4
Polymorphic Haplotypes inherited *en bloc*
Mating between the first and second subjects shown in Figure 3. On the left the alleles segregate independently whereas on the right the four haplotypes have segregated *en bloc*.

Interestingly, inheritance of "Linkage Groups" was recognised in Drosophila soon after the free recombination model became doctrine. Somehow, the implications were never incorporated into the precepts of population genetics. In fact, as shown below, the explanation for this surprising omission may be best explained by the demand for simplicity in the routines which survived the "bean bag" debates. "Genes" were defined by what they did (rather than what they were) and, without doubt included "linkage groups" as single genes for the purpose of calculations.

The major problem for population genetics has arisen from attempts to relate detection of genes to function or, as currently, to relate DNA sequences, including Single Nucleotide Polymorphisms, to phenotypes. The issue is summarised in Figure 4. On the left, the logic is to detect the independent alleles. On the right, the challenge is to track the haplotypes.

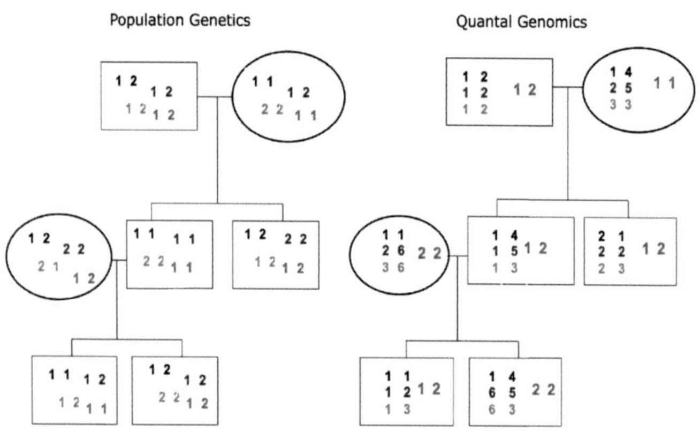

Figure 5
Stable segregation of ancestral haplotypes
Mating between a second generation member and subject #3 in Figure 3. On the left the alleles segregate independently. On the right the four ancestral haplotypes have segregated predictably. Note the introduction on the right of new polymorphic blocks in a parent for the third generation so as to emphasise the greater conserved polymorphism within blocks.

In Figure 5, we show that these same polymorphic functional haplotypes are protected from mutation and recombination. Once established, ancestral haplotypes are conserved. Thus, inheritance is not from immediate parents but from the ancestors of a species or breed or line, however defined. Phenotypic differences are explained by the faithful transmission of conserved ancestral haplotypes. On the left, each generation has a new set of genes of unknown significance. On the right, the phenotype is constrained and determined by whichever ancestral haplotypes have been conserved over multiple generations.

Figure 5 also shows the importance of at least three generations in defining ancestral haplotypes by direct observation of pedigrees. It is immediately obvious that at least three generations are essential to identify ancestral haplotypes. So called trios of only one mating are misleading. If a haplotype is transmitted through at least two meiosis, it becomes a candidate which must then be observed over many more generations by examining more descendants. The preferred criteria for an ancestral haplotype include all of:

1. No recombination.

2. No mutation.

3. Identical indels.

4. Identical copy number.

5. Association with a function or trait.

In practice, however, some assignments will be provisional.

Figure 6
Spontaneous mutations?
Gene mutation indicated by asterix. On the right the mutation will have little or any consequence because truly important differences are encoded within the conserved block. Monogenic diseases or traits are the partial exceptions. On the left they may or may not be important, depending upon frequency, context, repair and heritability.

The genomic distribution of mutations is illustrated in Figure 6. Since mutations are suppressed within ancestral haplotypes, they can be expected in intervening essentially non-polymorphic sequences. This is the case. It follows that detection of mutations will be of minor significance to breeders.

This figure raises the interesting paradox: Population genetics is based on unimportant mutations which follow the rules of the infinitesimal model.

Predicament Facing Breeders

Here we offer some possible explanations for the predicament facing breeders with particular reference to dairy and beef breeders. Should they stay with EBVs - including so called genomic EBVs - in order to meet the market? Is there an alternative?

Oversimplification

At the simplest level, the "modern" approach has been reductionist rather than holistic. It was always overly optimistic to expect that continuous variables reflecting complex traits would behave like a Mendelian character or like a monogenic trait or disorder. More recently, some seem to expect that complex traits will be marked by a "single nucleotide polymorphism" or some combination of SNPs! To quote Meuwissen and Goddard [7]. "assuming that the causal SNPs are included … the analysis simplifies … because the effects of all other SNPs … are expected to disappear". By whatever logic, there were attempts to separate critical functions like health, milk production, fertility as if they could be manipulated independently. Clearly, in the long term, each depends on the other.

Momentum of old arguments

A century ago, the "Mendelians" believed that discrete segregating genes could explain continuous variables whereas the Galton "biometricians" argued, in effect, that inheritance was more complex. The Mendelians appeared to prevail and, following R. A. Fisher, eventually gave birth to the new discipline of quantitative or population genetics [10]. More recently, a whole profession has been built on analysis of Quantitative Trait Loci (QTL) and Estimated Breeding Values (EBVs). Amazingly, old Mendelian concepts continue to prevail even though advances in genomics support the arguments of the biometricians. In fact, as is so often the case when arguments become polarised, both sides were correct to some degree. The Mendelians can relate to simple traits and diseases, such as monogenic protein deficiencies, whilst the biometricians anticipated very complex traits and diseases such as milk production and diabetes. Practitioners of artificial selection need a new approach incorporating both schools but based on the current understanding of the vertebrate genome.

Premature commercialization

Following the widespread use of Artificial Insemination and then Embryo Transfer, semen became a marketable commodity. Promotions encouraged claims of superiority based on limited data often presented in a simplistic almost contrived way. Indices and manipulated values are provided without adequate explanation. Sometimes, it seems to be forgotten that semen is a haploid mix and a poor representation of *cis*, *trans* and epistatic interactions.

Diagnostic companies have offered tests for a range of QTLs, often without statistical justification or independent validation. For example, van Eenennaam et al [11] revealed that commercial claims were unjustified.

SNP typing was packaged for all to use and promoted as valuable in selection even when predictive values were very low and even when, as illustrated below, the results indicated the need for an approach based on genomics and ancestral haplotypes. Unfortunately, the momentum of commercialization has come to dominate selective breeding.

Private data

Frequently, data used to promote commercial products are not available for independent assessment. Secrecy has become acceptable. Adoption of scientific traditions such as disclosure and peer review are overdue and could have avoided some of the adverse reaction now occurring. It appears that reasonable standards may have to be enforced by regulation.

Isolation

Although livestock breeders are conducting critical experiments in evolution, they are doing so without considering issues which are now mainstream in vertebrate molecular genetics and genomics. For example, how do duplications, indels and paralogous regions contribute to inheritance? How is expression regulated? What is the role of polymorphic blocks? These questions are critical and should be the focus of animal breeders with the added incentive that answers will benefit all vertebrates and especially humans. Indeed there are real opportunities as it becomes clear that classical studies in plants, flies and worms can be very misleading and as, at the same time, humans surround themselves with disincentives to investigate themselves. Genomic analysis shows that Bota and Hosa have many close paralogous similarities waiting to be exploited.

Potential of Comparative Genomics

Genomic analysis has confirmed Darwin's suggestion that evolution has developed by successive events of speciation akin to the branching of a tree. The new species retains most of the preceding genome but with species-specific variations. Cellular structure and function are largely conserved through each speciation event. The amount, distribution and regulation of cell products, such as protein, can change substantially whilst the actual sequence remains unaffected.

Table 1 Evolution of Animal Species

Species	Last Common Ancestor with Human (mya)*
Earth worms (*C.elegans*)	938
Fruit flies (Drosophila)	783
Tiger shark	462
Bony fish	400
Birds	302
Kangaroo	163
Mice, rats	92
Cattle(ungulates)	94
Dog	94
Lemur	74
Macaque	29
Chimpanzee	6

* Millions of years go, mya. Last Common Ancestor from mean of collected results at http://www.timetree.org/

The time scale of the speciation events is very largely agreed even though estimations have been based on different assumptions and measurements. A summary is given in Table 1.

The separation time between Hosa and Bota is similar to that between primates and quite small in relation to the entire vertebrate lineage. In Figure 7A, we show that the cow, Bota, has large stretches of genome which are remarkably similar to Hosa. The main changes involve:

- shuffling of chromosomal segments including transpositions and inversions,
- copy number.

As shown in all three dot plots, but most dramatically in 7G, the homologous sequence is almost continuous through the paler squares - corresponding to coding regions - and the darker hatched squares - corresponding to introns and intergenic regions. The greater background density is due to the concentration of retroviral like elements and related regulatory sequences. Careful inspection reveals the presence of indels. In 7C and 7E the homology is reduced and the indels are larger resulting in interruptions and deviations. The locations shown in red

---

Figure 7
Human Chromosome 17 (HSA17) *versus* Bovine Chromosome 19 (BTA19) homology

Figure 7A Selected loci (arrows) potentially involved in horn or muscle development or fatty acid metabolism.

MYO19, MYO1C, MYO1D: Myosin related
ALOX15, ALOX12 Lipoxygenases. Reported scurs locus region (Asai Animal Genetics v35pp34 2004)
SREBF1:Sterol regulatory element binding transcription factor 1
MPRIP: Myosin phosphatase Rho interacting protein
TCAP: Titin-cap (telethonin)
KRT: Keratin gene region
GH1: Growth hormone 1
FASN: Fatty acid synthase
BTA19: Bovine chromosome 19
HSA17: Human chromosome 17

Coloured boxes represent orthologous segments.
Numbers represent Mb.

Reference gene names and locations were taken from genome.ucsc.edu. Human sequence, Dec. 2013 (GRCh 38/hg38) Assembly; Bovine sequence, Oct. 2011 (Baylor Btau_4.6.1/bosTau7) Assembly.

Blue lines represent orthologous regions translocated.

Red line represent orthologous regions translocated and inverted.

In the dot-plots of Figures 7 C,E,G Hosa sequence is along the horizontal axis and the BTA sequence is along the vertical.

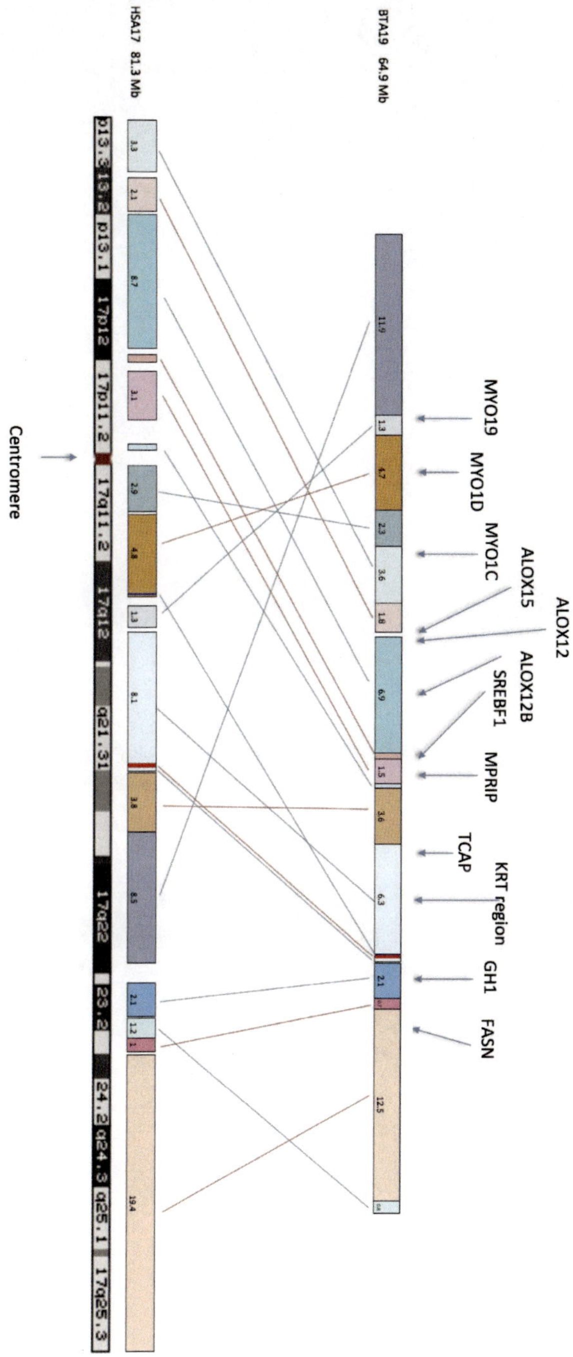
HSA17 81.3 Mb
BTA19 64.9 Mb
Centromere
MYO19
MYO1D
MYO1C
ALOX15
ALOX12
ALOX12B
SREBF1
MPRIP
TCAP
KRT region
GH1
FASN

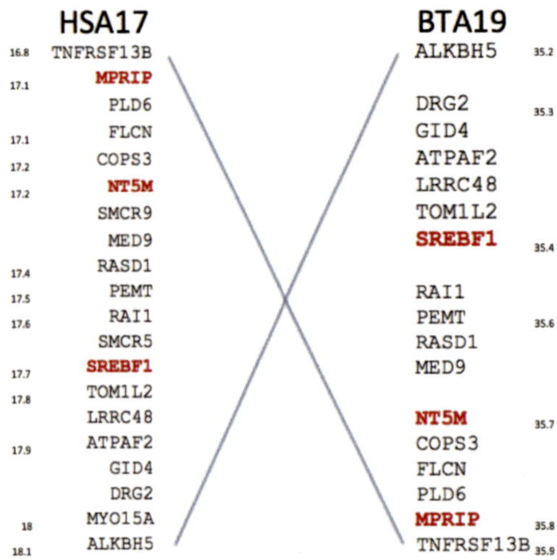

DOTPLOT analysis of HOSA17 18.8-18.1Mb (17p11.2) and BTA19 35.2-36Mb

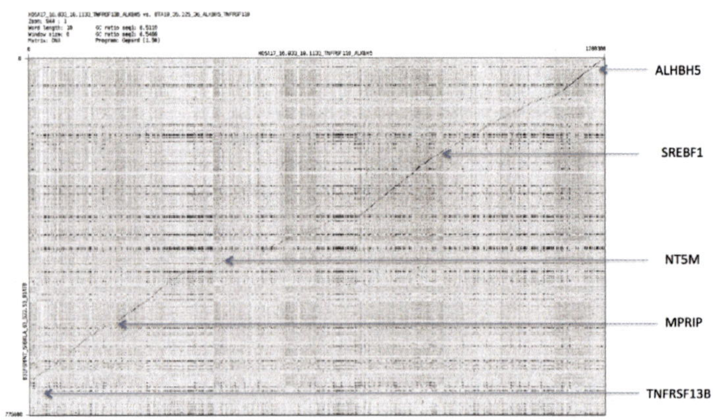

Fig. 7B-7C

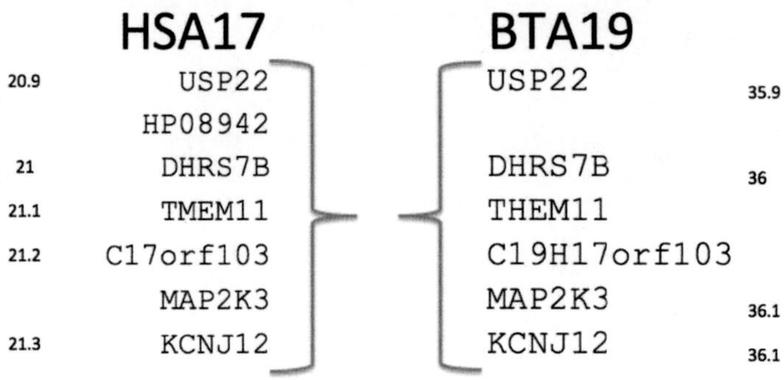

DOTPLOT analysis of HOSA17 21.1-18.1Mb (17p11.2) and BTA19 36-36.2Mb

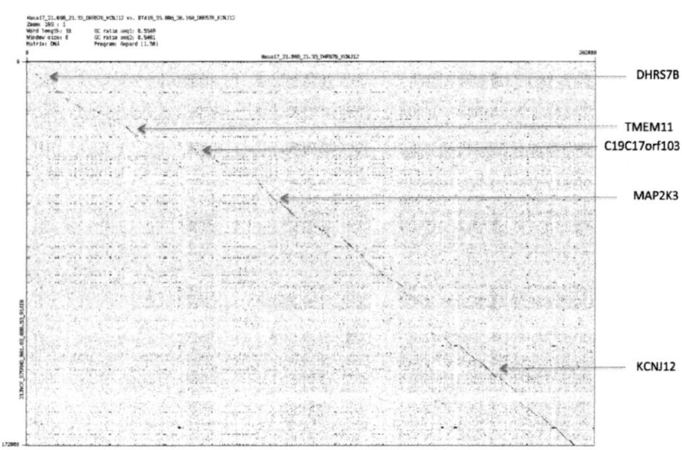

Fig. 7D-7E

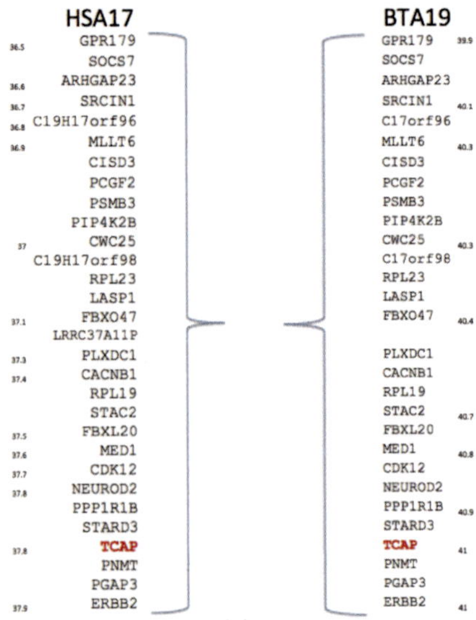

DOTPLOT alignment shows identity of genomic DNA between BTAC19 and HosaC17 in a 1.4Mb region containing TCAP

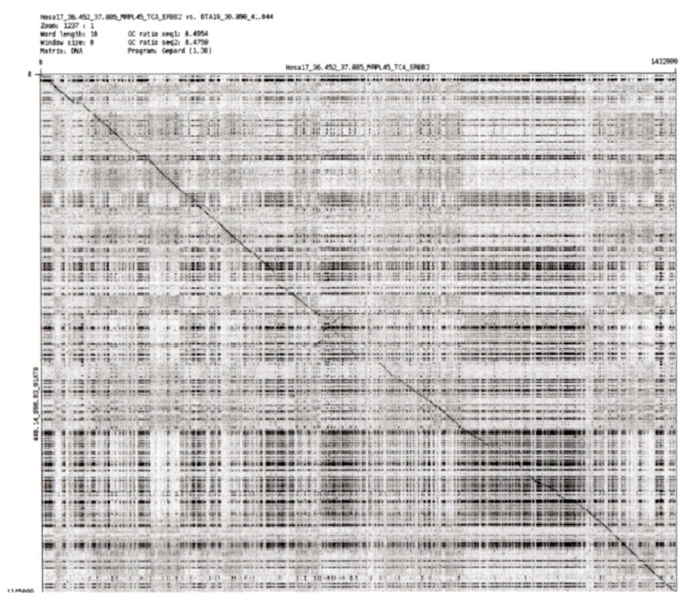

Fig 7F-7G

are used to define ancestral haplotypes in all breeds of cattle (see below).

Implications from such analyses include:

- Human genomics provides a useful guide to the equivalents in livestock.

- Sequence homology relates to noncoding as well as coding elements.

- Conservation has been maintained for millions of years and therefore many thousands of generations and meiosis.

- Architectural changes such as transpositions and copy numbers are more promising than seeking individual base changes or SNPs.

- SNPs will fail to detect major changes in genomic structure and gene regulation.

- Ancestral haplotypes are useful because they reflect all genomic differences.

The Genomic Approach

The approach proposed here depends upon an understanding of genomic evolution rather than population genetics (see Figure 8). It takes account of:

- Ancestral haplotypes (ie extensive conserved polymorphic sequences) as the unit of inheritance.

- Variable penetrance.

- Copy number and paralogous redundancy.

- The critical importance of noncoding and regulatory sequences.

- The coevolution of man, his food and their shared parasites including retroviruses,

and, most importantly.

- The concept of anamnestic evolution based on long-term conservation and blending of ancestral haplotypes.

The forces of natural selection and evolution are engaged rather than ignored. The power of meiotic recombination is exploited actively.

A strategy follows. This requires the breeder to study both parents plus their families and ancestors rather than just the sire and the grand sire. Effectively, the information required is carried by the diploid mixes of ancestral haplotypes rather than so called "mutations" or "single polymorphisms". Therefore judgments can only be made after comparisons over multiple generations rather than sibships and single matings. The aim is to breed a genome, which incorporates a deep memory, enriched for the complex functions required by the breeder. Secondary benefits of the approach include direct haplotyping which permits sire and "breed" determination, traceability and food safety.

Selective Breeding In Livestock

Undoubtedly, classical breeding of livestock has been very successful in adapting animals to environments and in improving commercial outcomes. How?

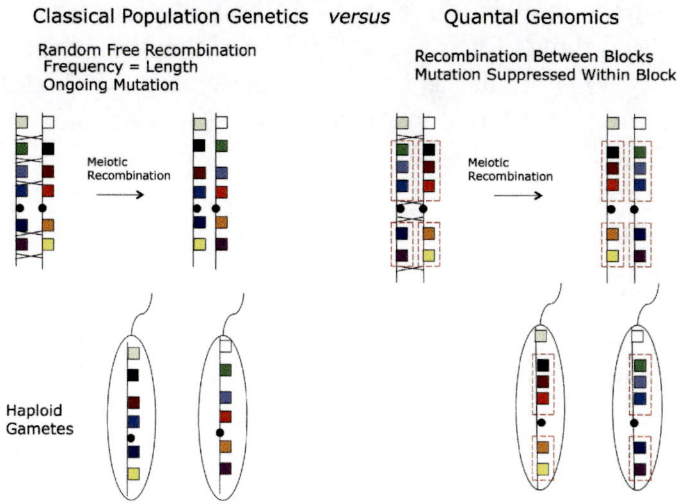

Figure 8.
Population Genetics *versus* Quantal Genomics

Prior to genetic testing, the selection was based on phenotypes and direct observations of the mode of inheritance. Although there are well known examples of dominant or recessive modes, most traits were intermediate and complex meaning that, in practical terms, desired change required multiple generations.

Until quite recently, most breeders depended entirely upon their own experience and discussion with fellow breeders. Although informal, the database was huge and powerful given its depth in time. The result was judicious and holistic blending.

These concepts are illustrated in a review of the development of Dorset sheep in South Australia during the 43 years from 1917 to 1960.

Thus, W. J. Dawkins personal communication (1960) writes :

"It is necessary to keep careful records of the mating time, birth, tattooing, and earmarking for sires to make possible the immediate identification of each lamb, and to ear-tag to associate sire and dam to the offspring. Strange to say, when I went to England with my eyes and ears open to strive to learn from these old breeders how they did things I was astounded to find that they were not nearly so particular as I was with these very essential details, and even found that many were using two rams with the one flock, so that they could not give exact details of the sires, but put in the pedigree such things as 'by either one ram or another', something that I could never bring myself to do.

"Now the Poll Dorsets are setting up a very fine record and the Poll Shorthorns are causing strong interest in these same shows. It is interesting to note that I have exported sheep to all States of the Commonwealth, to South Africa, New Zealand, England, and to America. My type has changed since I first started breeding Dorsets, for at the start they were much taller, lacked meat in the twist, possessed the Devil's grip somewhat, but we breeders of South Australia have filled their twist, widened the loin, eliminated the Devil's grip largely, shortened their leg so that today the Dorset is the most popular breed for the fat lamb industry in Australia, all being done in the comparatively short time of 43 years, and S.A. is now known as the home of the Dorset in Australia, and in all this I took a very active part, both in the registration society and in the type developed.

"Another day I remember Ross Grant, the Chief Veterinary Officer of the Board opposing any other lamb being graded as Down unless they were bred by Southdown rams, and once went so far as to say he had instructed his graders to look in the live pens before slaughter to estimate the number of lambs with the brown or black legs which he said was indicative of the down sire, and that they were not to grade any more than the estimated number of Downs when they saw them on the hook, than those with the black legs. I also remember another lamb buyer at Smithfield London telling me that we Australians should leave a little of the wool on the leg so that the buyers could be sure that they were Downs. To both of these ideas I took very strong exception, and said some very outspoken words at the Board in Grant's presence. I said that if the grader could not tell what sort of a lamb it was without seeing the wool on the legs he was not worthy to be a grader; that the dressed carcase was the final test, and if other breeds could produce as good a lamb as the Southdown then they should be graded with equal standards. I think from this advocacy that the Down TYPE was established, a term now fully recognised."

Clearly the attitude at the time was to select for type (multiple interacting traits) determined visually and by carcase. Subsequently the emphasis moved to discrete variables amenable to statistical analysis. Now as the wheel turns, there is renewed interest in TYPE at least when EBVs and the studbook are unhelpful.

Domestication and Development of Breeds

The selection processes which have resulted in the livestock of today were truly effective, even if intuitive. The extent of phenotypic change over a few hundred, let alone a few thousand years, is striking when measured against similar species in the wild or even the same domesticated species after natural rather than controlled mating. Surprisingly, reversion to the wild does not appear to have been studied systematically. However, it is said to occur quite rapidly (Figure 9), as for example, when the preferred, selected sire is less fertile than the natural competitor. Apparently, the transience of a selected phenotype is due to dilution within the ancestral genome, illustrating the depth and power of the natural selection which resulted in those ancestors. In this context, the success of any form of controlled selection is surprising and requires explanation. Certainly, it could not have been predicted from population genetics or any evolutionary teaching based on the selection of mutations. To the contrary, the explanation appears to be more in keeping with the idea of progressive blending.

Figure 9 Reversion to unselected wild in Vanuatu results in dog-size cattle. (Photo by the author)

Although domestication can be regarded as one of the most informative experiments in evolution, the experimental design was not ideal. The aims must have been as varied as:

- Tractable behaviour.
- Safe and fresh food including milk.
- Ability to reproduce.
- Freedom from transmissible disease.
- Capacity for work.

The strategy for selection must have been comprehensive rather than trait specific. Presumably, major phenotypic changes were avoided, especially if there were advantages and disadvantages as seen more recently when one aim dominates over others. Geographic isolation may have helped to consolidate the "gene pool" resulting, eventually, in more predictable outcomes within that breed. If so, there would have been different populations selected for similar purposes. Eventually, these populations would be mixed allowing genetic blending and further consolidation of desirable haplotypes and dilution of breed specific disadvantages.

A further aspect of the domestication experiment is that the endpoints were blunt but biological. Whilst a disadvantage in some respects, such endpoints would have resulted in a cumulative and careful approach and, consequently, may help to explain the success. At the least, there were no commercial incentives for rapid decisions based on pooling of statistical indices.

Simple Traits Are Not So Simple

Valuable lessons arose from experience through selection for apparently simple traits.

Polling in cattle can be taken as an example of an apparently dominant character (PP or Ph). Horning is recessive (hh). A bull was designated heterozygous (Ph) if any of his calves were horned. Polling could be traced back through many generations implying heritability. Recent mutation was rarely if ever the explanation although such claims continue to appear. Evidence for what some call a "linkage group", but is more likely an ancestral haplotype, include associations between polling and other features such as corkscrew penis or abnormal prepuce [12] in bulls and poor milking in cows. Association with an eyelash-eyelid phenotype has also been suggested [12] (example Figure 10). Taken together, it appears clear that those seeking a single "mutation" in a coding sequence will be disappointed, as has been the case. Indeed, there is an alarming discordance in publications and patents relating to polling. Yet again, there is evidence of poor science combined with premature commercialization; it is not surprising that breeders have lost confidence.

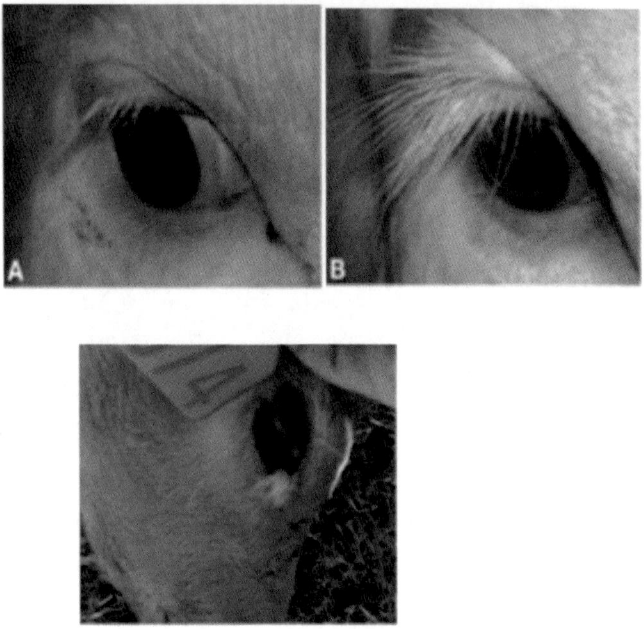

Figure 10
Horned animal A has single eyelash while polled animals B and below have thicker and irregular eyelashes.
See also Allais-Bonnet A, Grohs C, Medugorac I, Krebs S, Djari A, et al. (2013) Novel Insights into the Bovine Polled Phenotype and Horn Ontogenesis in Bovidae. PLoS ONE 8 (5) : e63512. doi: 10.1371/journal.pone.0063512

An understanding of the impressions of older breeders should help to redress the situation. Unfortunately, much of this wisdom has been lost but some themes may be informative.

A homozygous bull could be produced by breeding "within the line" such that each of two descendants of a recent common ancestor contribute a haplotype identical by descent. The complication of concentrating undesirable hitchhiking genes could be avoided, in some cases, by crossing two distinct and remote ancestries so that each copy of the polling gene had a different ancestry and therefore differences in the accompanying genes

However, it was thoroughly appreciated that the predictability of polling decreased as the two ancestries diverged, suggesting that modifying genes were involved and that the precise content of each carrier haplotype may be partly breed specific. For example, it is widely believed that polling through the Angus "gene" is less predictable in Angus crossbreds than in pure Angus. In order to develop Polled Dorsett sheep it was necessary to "consolidate" the Corriedale "gene" before crossing heterozygotes to produce homozygotes. By extrapolation from what is now known of vertebrate genomics, such haplotypes are likely to differ in copy number, indels and non-coding regions affecting expression.

If homozygosity could only be achieved reliably through sharing of immediate ancestry, such as brother-sister mating, diverse recessives were likely to be expressed. In this case, the other genes need not be a component of the carrier haplotype and, of course, need not be linked to polling haplotypes.

Earlier breeders were very conscious of the complexity of penetrance. For example, horns appear at various ages. When the head is square rather than peaked, horns can be expected even if not palpable at birth. In some cases, the apparently polled calf will develop "scurs", defined as mobile horns. Over time the scurs can become fixed but the resulting horns are irregular rather than smooth and shiny, or vestigial (see Figure 11). This scurring "gene" appears to be unlinked but is only expressed in the presence of incomplete polling. If, as has been suggested, polling is controlled by haplotypes on BTA 1 whereas scurring haplotypes are on BTA 19 [13] or BTA 4 [14], the dependence of the second on the first implies a far from simple form of epistasis.

In the case of sheep, the impression of a single dominant was replaced with a far more complex model. Thus,

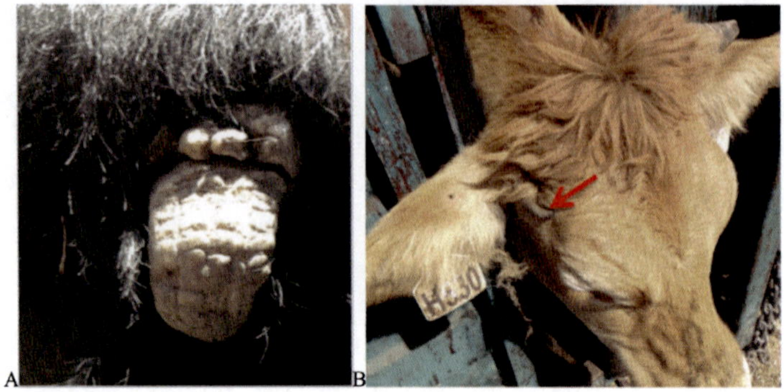

Figure 11
Scurs in cattle
A) Serrated mobile scur illustrates that the genetic regulation is aberrant. B) Normal Left Horn and Vestigal Right Horn/Scur. Note central hair pattern. The red arrow indicates the vestigial right horn

returning to Poll Dorsets, the observations are very clear. To quote the official history:

"The first real attempt to poll the Dorset Horn was begun by Mr W.J. Dawkins, veteran breeder of Newbold, who had done so much over 40 years to adapt the breed to Australian conditions. In 1937, a stud Corriedale was accidentally mated to a Newbold stud Dorset Horn ewe and the resulting progeny was a poll ewe lamb. Mr Dawkins for some time had been considering a Poll Dorset experiment, so this happening gave his idea the necessary impetus.

"Mr Dawkins, together with Mr J.W. Reddin, who handled the Newbold sheep, believed it was most important in the selection of foundation stock for such a venture that the sheep should be of good carcase quality. This applied equally to the pure stock used from year to year during the grading up process. They felt that if this point of good carcase quality was watched from the start, there should be no necessity to reintroduce horned blood into an established poll flock to maintain quality. In fact it was soon established that top quality ewes were vital for sound progress.

"The establishment of the Newbold Poll Dorset flock was planned in three stages. Firstly, the policy was to mate poll rams on pure ewes, and pure rams on poll ewes, year after year until at least 96.875 per cent pure Dorset Horn blood was attained. In actual fact, most of the Newbold Polls were taken initially to the 98.4 per cent stage.

"In the second phase the aim was to build up a poll flock with double poll parentage by mating polls on polls of at least 96.875 per cent purity.

"In the third stage the objective was to locate dominant polls by test-mating and so form the nucleus of a true poll flock and provide other breeders with rams for mating with pure Dorset Horn ewes.

"The first aim was to fix the inherent Dorset Horn characters, and the second and the third to fix the poll character. The Newbold experience was that any attempt to short-cut the programme by mating polls on polls too early could prove disastrous from the genetic point of view of the Dorset Horn.

"Mating back to the Dorset Horn required heavy annual culling for horn growth. This was discouraging and impatience led in 1947 to a mating of polls on polls with the objective of breeding a double poll ram to use more effectively on the pure ewes in the grading up process. Progeny showed retrogression, which meant more culling and waste of time and sheep.

"These experiences brought important lessons. With the constant need for more poll ewes, the importance of an obvious 'sex' factor in the breeding of Poll Dorsets was not realised early enough. It was not only much easier to breed the horns off the ewe lamb than the ram lamb, but it was also a fact that the use of the pure Dorset Horn ram on poll ewes would not bring anywhere near the same results as the use of the Poll ram on the pure Dorset Horn ewe. The whole experiment could have been considerably accelerated by using more pure-bred ewes. This was an expensive discovery, as the ruling price of pure Dorset Horn ewes was then high.

"Once the grading up stage had been completed it was found that the culling rate dropped enormously. Another encouraging feature was that the sex factor was steadily eliminated, with ram and ewe progeny being polled alike. The whole poll percentage increased steadily with the use of double poll rams and by progeny testing to find the real dehorners of quality.

" While the Newbold studmasters used the Corriedale to achieve the poll factor, L.A. Stuart and R.H. Wilson introduced the Ryeland.

"Mr. Stuart began his project in 1941 with the purchase of stud Ryeland ewes which were joined to a top stud Dorset Horn ram. Mr. Stuart followed that policy throughout the grading up years, always using the stud Dorset Horn ram.

"By doing this he felt he would have less variability than if he used a poll crossbred whose selection would have to depend to a large extent on his poll characteristics, rather than general Dorset quality. By the time the 96.875 per cent percentage was attained it was found that the Dorset Horn rams were siring nearly all horned ewes and no true poll rams.

"At that point it was recalled that one Dorset Horn ram had proved good for polls. He was repurchased and sired nearly all poll ewes and some really good poll rams. From that time, the Valma project was on a sound basis.

"Mr. R.H. Wilson began his experiment in 1943, when he initially introduced Ryeland rams and then subsequently bred poll rams to Dorset Horn ewes, and Dorset Horn rams to poll ewes. The initial ewes were high quality cast-for-age ewes from his own Dorset Horn stud.

"As with the other breeders, he found that progress was slow at first because of the limited field of selection, so he allotted larger numbers of Dorset Horn ewes to the project.

"This involved the purchase of two additional Ryeland rams and it was found that great care had to be taken in their selection. Fortunately, one proved outstanding for quality. The other had to be discarded. As with

Newbold and Valma, real progress came only with a ram dominant for the poll factor.

"With the exception of Temora which had a Ryeland infusion, all other early flocks drew a poll ram from one of the three pioneer breeders, and mated them to Dorset Horn ewes. This was the obvious short-cut to the establishment of a Poll Dorset flock.

"These three men had to withstand much criticism. They were accused of breeding crossbreds, and there was even talk that they might be expelled from the ASBBS for it. However, geneticists of world standing agreed that it did not matter a whit which poll breed was used in the first place provided that the back-mating programme was strictly adhered to until a sufficient percentage of pure Dorset Horn blood was attained.

"All three pioneer breeders were adamant about the back-mating programme – hence the insistence of a minimum of 95 per cent of Dorset Horn blood for registration. In fact, they considered that the closer breeders got to the 100 per cent of Dorset Horn blood the better. There would be few sheep, if any, in the breed today that do not carry 99 per cent of Dorset blood.

" By 1954, after years of careful selection and ruthless and costly culling, Messrs. Dawkins, Stuart and Wilson felt that they had produced Poll Dorsets of sufficient Dorset Horn quality to establish the Poll Dorset breed of sheep." (www.polldorset.org.au, with permission).

In 1964 C. H. S. Dolling produced a breeding manual at the request of the Poll Dorset Association. Although Dolling used the concepts of Fisher, Lush and Falconer he clearly understood that it was important to track through generations and to consider interactions affecting phenotypes.

Experience in vertebrate genomics teaches that more than one chromosome will be involved in all but the simplest of monogenic traits. This is because extensive duplication and translocation is a feature of vertebrates. In many examples, genomic analysis reveals that paralogous

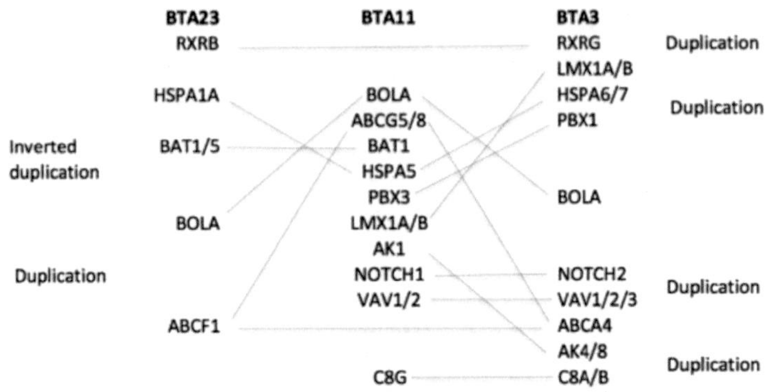

Figure 12
Paralogy between MHC on BTA23 and paralogous groups on BTA11 and BTA3.
The genes are listed in order from the telomere. The arrangement can be largely explained by duplication with and without inversion events. ABC - ATP-binding cassette; AK - adenylate kinase; BAT - HLA-B associated transcript; BOLA - MHC class I, C8 - complement component C8; HSPA - heat shock protein, LMX1 - LIM homeobox transcription factor; NOTCH - neurogenic locus notch homologue; PBX - pre-B-cell leukemia transcription factor; VAV - guanine nucleotide exchange factor.

clusters have been replicated and transposed before diversifying. In this way, redundancy is available with the benefit of protection against dependence on a single location. Further, in keeping with many other examples, redundancy leads to the development of polymorphic interactive pathways including sequential cascades. It seems highly likely that horn development will prove to be another example of control through paralogous clusters containing polymorphic frozen blocks encoding numerous ancestral haplotypes. A testable hypothesis is that polling and scurring are controlled by interactive paralogous pathways affecting expression of keratin. Indeed there are paralogous regions for the scurring QTL on Bota 19 and the polling QTL on BTA 1 (not shown). Another example of clear paralogy is between the bovine MHC region on BTA23 and paralogous groups on BTA11 and BTA3 (Figure 12).

These features can be summarized in an attempt to assist future investigations:

- Polling may appear to be a single dominant gene but other linked and perhaps unlinked genes are involved.

- Scurring is also complicated with degrees of expression in multiple tissue types.

- Each genotype affects the shape of the skull and must be pleiotropic.

- The responsible haplotypes are ancestral (con served extensive polymorphic sequences) rather than simple carriers of a recent mutation.

- These ancestral haplotypes may be expressed in diverse pathways and may differ qualitatively and quantitatively in different breeds.

- Penetrance is variable with many, yet to be understood, modifiers.

- Genetic interactions (*cis*, *trans* and epistatic) need to be elucidated before the genetics will be entirely clear - even though artificial selection has been so effective in producing polled breeds over countless generations.

- Paralogous regions will be involved.

- The consequences of inbreeding are difficult to predict.

These few points raise some interesting issues which could be addressed with spectacular benefit to vertebrate - including human - genomics. They raise fundamental issues if the genotype-to-phenotype interactions are to be understood. Specifically they suggest that inherited polymorphisms can have remote effects which elsewhere we refer to as the Multifunctional Polymorphic Controller (MPC). The key idea is that polymorphisms can interact with each other such that a given diploid heterozygous individual has a range of outcomes much as the cogs of a gearbox allow different speeds (see Chapter 4). Thus the individual possesses an inherent capacity to adapt to circumstances including changes in internal and external environments. It follows that biological pathways may be less predictable but more robust through heterozygous advantage. A further implication is that there can be stochastic differences between bilateral versions of polymorphic characters such as horns. Thus right and left scurs should not be expected to be identical (see Fig 11).

Similar conclusions are possible for other traits such as coat colour. The consistency and complexity of the inheritance is immediately apparent although, in this case, further compounded by the addition of quantitative variation attributed, in the past, to intrauterine epigenetics and melanocyte migration (Figure 13). We prefer stochastic meshing due to heterozygous regulation. Such quantitative variation must also apply to less apparent phenotypes.

Another example of relatively obvious inheritance is "double muscling" as found in Belgian Blue cattle (Figure 14). Although apparently simple, other characteristics segregate with muscle phenotype. Forms of double muscling can be found in other breeds and can be readily distinguished. Different haplotypes are involved even though, in the case of the myostatin gene, the biochemical effect of different polymorphisms and impaired expression may be largely the same.

A study by Grobet et al. [15] is informative in revealing that coding polymorphisms may be over interpreted. Myostatin sequences in ten breeds were compared. Of the eleven sequence polymorphisms, five were predicted to disrupt function but were not present in two of the ten

Figure 13
Cloned animals and identical twins show very similar, but not identical markings.
Differences are said to be due to intrauterine melanocyte migration. (Photograph by Dr Mal Brandon with permission)

breeds suggesting either "additional mutations in unexplored segments of the gene or…locus heterogeneity". A preferred approach would be analysis of all coding and noncoding sequences with determination of haplotypes.

Complex Traits Are Very Complex

The examples of apparently simple traits illustrate the inherent complexity of vertebrate genetics, which, however, is hugely compounded for those traits which are clearly polygenic and obviously very dependent upon environmental factors. Unfortunately, these so called quantitative traits are the most important commercially but the most difficult to manipulate. Calving ease, fertility, milk production and feed conversion are quantitative traits, which are critical to efficient production. Some authorities seem to believe that the importance of these characters has only been appreciated recently and that improvements will be immediately possible through the application of quantitative genetics.

Figure 14
Double muscling involves different polymorphisms in different breeds

A) Double muscling in Belgian Blue bull with a deletion in the third exon of the myostatin gene, which reveals a premature stop codon.(Photograph from agriflanders http://www.flickr.com/photos/70614216@N00/3187975982)

B) Charolais bull, with a Q204X C to T substitution in the second exon of myostatin yielding a premature stop codon. Australian Charolais Bull at the Melbourne show 2005 photograph by Fir0002/Flagstaffotos http://commons.wikimedia.org/wiki/File:Charolais.jpg.

C) Highly muscled Limousin have the F94L polymorphism which would not be expected to cause dysfunction in the myostatin gene. photo from http://commons.wikimedia.org/wiki/File:BuckinghamshireCo.show.2006.gif

D)Holstein Bull without double muscling. http://www.licnz.com/holstein_friesian.cfm

It is often overlooked that these complex traits have been understood in practical ways and already manipulated to a remarkable degree. Continuing improvements are just some of the motivations which have resulted in the evolution of the various breeds seen today after thousands of years of domestication and at least 200 years of pragmatic selective breeding. It is clear that the consequence was the development of identifiable breeds with all critical traits to various degrees. Perhaps it was the pursuit of multiple features which itself explains the success of those breeders.

Once established these distinctive breeds were crossed with each other to produce improvements in some characteristics or "heterosis" or "hybrid vigour". In retrospect, these improvements could only have been predictable if the cross were between distinct breeds which themselves, over time, had bred predictably. The early breeders talked of blending the different types recognizing, as Darwin did, that the process was complex and the result often surprising including beyond expectation. Poultry and pig breeders seem to have learnt these lessons.

The Example of Wagyu

The history of Wagyu cattle is informative. Apparently native breeds were crossed with British and European breeds to produce extreme marbling to a degree not seen in the ancestral populations. It would be very useful to know just how this was achieved. There are several versions of the story. One favourite is that the breed began from crossing local cattle around Nagasaki and Hiroshima. It seems plausible that the crosses included Jersey dairy cows introduced by the Portuguese sailors in the late 17th century. According to Namikawa [16] (cited by David Blackmore) there was little interest in milk or meat until 1868 and very little infusion until about the same time. There had however been inbreeding of the native cattle from at least 1830 with the creation of the Tsuru with desired characteristics such as healthiness, fertility and mild disposition. Apparently, with the partial exception of the native cattle on the island of Mishima, most of these Tsuru have been lost through cross breeding but it is possible that some of the desirable features, such as excellent meat quality, have been preserved.

Cross breeding was encouraged between 1868 and 1910 when the focus changed to improving the native cattle, especially with respect to milk production. It seems, however, that these efforts were geographically isolated and that there were wide variations, some of which may have been related to earlier crossing with different foreign breeds. For example, Japanese Poll in Yamaguchi are thought to reflect Angus, the Japanese Brown in Kochi and Kumamoto particularly Simmental and the Japanese Shorthorn in Aomori, Iwate and Akita some mix of shorthorn, Devon and Ayrshire. The Japanese Black from different prefectures dominate numerically (85% of Wagyu in 1990) but are heterogeneous and trace back to infusions with Brown Swiss, Shorthorn, Simmental, Devon etc.

From 1950, in-breeding unions were established. After eliminating one strain with lethal or defective genes, the inbred lines were crossed systematically from 1962 to produce the Japanese Black of today. Experienced breeders were in charge. Rules included "cleanness of undesirable genes". The focus of different unions varied. For example, in the Hyogo prefecture, where, it is said, there was no introduction of new genes for 90 years, the cattle are characterised "by genetic excellence of beef quality and finely marbled Kobe or Matsuzaka beef."

Following the introduction of measurements, complexities became apparent such as:

- Excellent marbling of heifers and steers but low progeny testing of bulls.

- Unpredictable breeding values even when bulls are of the same strain.

An interpretation of the lessons from the Wagyu history might be as follows;

• The ancestral populations were selected for farm use rather than milk or meat.

• Geographic isolation closed these populations until recently.

• Over the past 150 years periodic inbreeding has alternated with blending with distinctly different populations including diverse imports.

• Recessive disorders have been troublesome but avoided as much as possible.

• Preferred characters are likely to be intermediate or recessive in inheritance.

• Multiple haplotypes are involved.

• Penetrance is multifactorial.

• There is a need for an intermediate degree of diversity.

At issue, in all breeds but especially Wagyu, is the best way of utilizing the vast experience of earlier breeders whilst adding the benefits arising from new technologies and concepts.

The introduction of Wagyu into Australia can be examined as an experiment in evolution. According to Dom Bayard (personal communication 2013), the use of different pure sires revealed unexpected and puzzling genetics. For example, Angus and Holstein cows inseminated with Michifuku of the Tajima line consistently produced marbled calves whereas many other equally prestigious bulls were total failures in this regard. Interestingly, however, some of these failures conferred high penetrance for other features such as size, milk and temperament. Thus, it appeared superficially that there was bull-specific dominant inheritance of diverse characters. With further experience, this explanation became less likely since the transmission from the sons of these bulls appeared to be much less than 50%.

Over time it became possible to compare bulls when used with 7/8, 15/16 Wagyu cows. In Figure 15 MRIP haplotypes (Williamson et al [38]) are given for reference (see below). Michifuku (30.1; 30.2), who had been so successful in producing marbling in 1/2 breds, was found to be a failure through lack of milking ability. By contrast, TF147 (30.4; 40.3) of the Shimane/Itozakura line is very successful in producing marbling of the calves of Tajima cows although he suffers in failing to transmit to his sons and in carrying CHS. TF 148 is known for low milk but a tendency to lay down subcutaneous rather than intramuscular fat. His female offspring produce marbled progeny when crossed with Hirashigetayasu (30.4; 40.2) progeny.

Whilst such crosses are still being evaluated, it has been difficult to show that any recent bulls are superior to the foundation sires.

Some tentative conclusions are achieving credibility:

- Complex genetics with many interactions creating the appearance of variable penetrance.

- Hitherto unpredictable inheritance requiring pragmatic observations.

- Haplotypes for either subcutaneous or intramuscular fat deposition will be defined.

- Multiple traits are necessary for successful commercial outcomes.

- Superior Japanese sires may benefit from a deep history reflecting the concentration of haplotypes of incomplete penetrance.

Figure 15 Wagyu Bulls used in Australia

A) Itoshigenami TF148 60.1; 30.4 Excellent marbling, small, low milk.

B) Itoshigefuji TF147 30.4;40.3 plenty of milk and size but very slow to marble (CHS carrier).

C) Michifuku 30.2; 30.1; reliable marbling in F1 progeny.

D) Hirashigetayasu J2351: 40.2; 30.3 mixed reputation good milk average size, sons marble over the right dams.

E) Haruki II 60.1; 30.2 popular sire with size, milk and marbling only downside is the variability of the marbling with large seam fat.

F) Itozurodoi TF151 30.4; 60.1 seen in the early days as the perfect all round sire to use and has formed the basis of many herds in Australia. Still used but semen supplies are tightly held now.

Note the MRIP haplotype codes, see Fig 17,18.

These conclusions suggest that there are immediate prospects for rational selection given appropriate haplotype markers. At issue is the best method of identification.

Lloyd et al [17] have shown that the fat melting temperature (Tm) can be measured efficiently with very clear differences between long fed Wagyu (<37°C) and short fed Simmental (>37°C). Subsequently, we have shown that these different Tms can be attributed, directly or indirectly, to ancestral haplotypes such as 60.1 and 30.4 in the black Wagyu. Based on these observations, we are examining the role of diploid combinations in the different breeds. We now understand that the haplotype frequencies are very different in Wagyu and Simmental (see "W" plots below) and that the challenge is to produce robust animals, lacking monogenic deficiencies, with the appropriate high frequency diploid combinations. Cattle breeders will need to recognise that "superior genetics" requires an understanding of deep ancestry and haplotype interactions.

The Example of Akaushi or Brown/Red Wagyu

Although often classified as Wagyu (or Japanese cow), Akaushi are a special case and readily distinguished from Black Wagyu or Kryoshi illustrated above.

Those now in Australia and the Americas are sometimes said to be descended from only 8 cows and 3 bulls which were airlifted to south Texas in 1994. Undoubtedly there were earlier imports and some contrived infusions. The colourful history of intrigue and litigation has yet to be told in crime novel form. Nevertheless, it does appear that the genetic pool has been very carefully maintained and that the progeny have remained true to type without any deleterious consequences. Presumably, the various breeders had a clear, albeit multi-trait, phenotype in mind and adopted well tried principles to maintain the whole rather than to improve particular traits at the expense of others. The consequence has been that all Akaushi whether in Japan, the Americas or Australia, retain defining features, such as:

- marbling on grass ,
- maturing slowly,
- retaining Japanese behaviours like trotting in file, kicking and sharing calves
- maintaining a pattern of solid brown,
- remarkable soundness in spite of "fine bone",
- continuing highly fertility in harsh environments.

The earlier history of the founder population appears to be non-contentious. Akaushi are descended from a very small indigenous family which grazed around Mt Aso in the Kumamoto region of the Kyushu island of Japan.

Population or Quantitative Genetics

There are several ways to define this discipline as reflected by the differences found in some of the undergraduate textbooks (a summary of the main contrasting points is displayed in Figure 2 through Figure 6). Gillespie says population genetics is "the genetic basis of evolution" with insights which are "theoretical rather than observational" and "not verified in a natural setting." [18].

Hartl and Clark offer definitions which are widely used by quantitative geneticists as distinct from others involved in genetics. "The science of population genetics deals with Mendel's laws and other... principles as they apply to entire populations..." and "...a polymorphic gene is one for which the most common allele... less than 0.95..." and "...a haplotype is the combination of alleles present on a chromosome" [19]. See also Dawkins et al. [20].

For the breeder,, the following features may be helpful approximations to the concepts used in livestock genetics:

- Discrete genes may have deleterious or benefitcial effects.

- Discrete genes can be identified by analysis of populations.

- Populations evolve by changing their gene frequencies.

- Genetic effects are due to changes in coding regions and are expressed in protein.

- Diversity tends to decrease by genetic drift.

- Heritability is related to variance or, more broadly, scatter.

- Evolution by natural selection requires ongoing mutation.

These concepts are consistent with "beanbag" genetics (p.18) involving modelling rather than observation and experiment. Nevertheless, the concepts have led to efforts to identify discrete genes and loci responsible for beneficial characteristics. At issue is why these ideas have become "all pervasive" [3], especially when incompatible with current understanding of the structure of the vertebrate genome (Figs 2-6).

Marker Assisted Selection

In a review of QTL in dairy cattle, Khatkar et al. [21], assessed genes/markers, or, at least, those in the public domain, for Marker Assisted Selection. The map shows

QTL on many chromosomes, designated here as BTA (Bos taurus autosome) 3, 6, 14 and 20, for example. Interestingly, milk, protein and fat yields appeared to be controlled by the same regions. Reproduction was also affected by multiple but different chromosomes such as BTA 1, 7, 10, 20, 21. Depending upon the measure used yet other chromosomes were involved including 23 near BoLA.

These results clearly illustrate the complexity of the genetics and the difficulty in partitioning the genes and their effects. In the hope of clarification, a meta-analysis was undertaken [21]. The 55 public domain studies were reduced to only 28 with sufficient information to be included. The conclusions were as expected from the earlier studies. There are QTL on 3, 6, 14 and 20 but also 1 and 9. There are at least two regions on 6; one near the casein complex affects milk production and another near BM143 also affects protein and fat yield and production. Interestingly, genomic analysis suggests that these regions are paralogous (Figure 16).

The authors [21] conclude that there is a need for finer mapping using more markers. As discussed above and below, the real need is for a different approach based on an understanding of vertebrate genomics. It is interesting that the authors do not cite a single paper on human or vertebrate genomics. Nor is there any reference to paralogy or the lessons to be taken from haplotype associations with disease and function. These lessons are important because they challenge the old dogma, adopted by Khatkar et al. [21], that closely linked markers in linkage disequilibrium (LD) should hold across populations and need not be re-established for each individual family. Experience in human genomics shows unequivocally that this strategy is invalid. Conserved Ancestral Haplotypes must be defined by direct observation of segregation over generations rather than predicted from microsatellites and SNPs. Only those haplotypes which are conserved are relevant to Marker Assisted Selection: if not, they cannot be useful markers. The identification of ancestral haplotypes is based on characteristic polymorphisms which include differences in sequence, indels or copy number; SNPs and microsatellites are misleading. The diversity of ancestral haplotypes is far greater than imagined by Mendelians. In fact, as shown by human studies [22, 23] summarised below, each individual has a particular diploid mix of haplotypes and is actually unique at the DNA level. In a sense, each individual has a unique memory of its previous generations and a vast experience available when required.

One way of applying the laws of genomics to milk production can be illustrated by considering the apparently negative effect of polling. Perhaps there are haplotypes

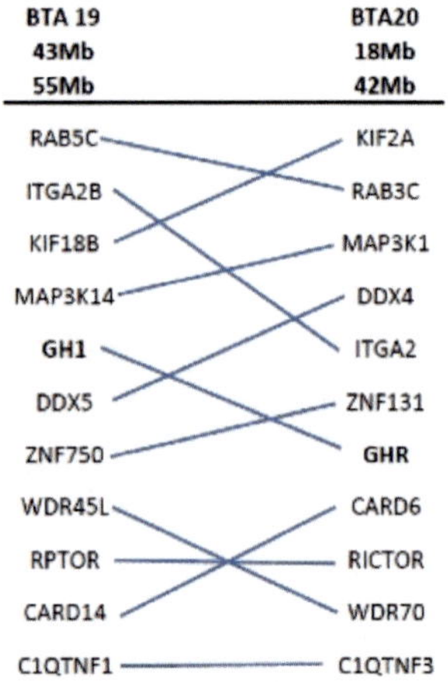

Figure 16
Interactions between Chr20 and Chr19

Potential paralogous locations in BTA19 and BTA20. The loci are listed in the order on the chromosome. Lines connect genes with similar function. The arrangements can be explained by duplication with and without inversion events. Note: C1QTNF, complement C1q tumor necrosis factor-related protein; CARD, caspase recruitment domain-containing protein; DDX, ATP-dependent RNA helicase; GH1, somatotropin; GHR, growth hormone receptor; ITGA, integrin alpha; KIF, kinesin-like protein; MAP3K, mitogen-activated protein kinase; RAB, ras-related protein; RICTOR, rapamycin-insensitive companion of mTOR; RPTOR, regulatory-associated protein of mTOR; WDR, WD repeat containing protein; ZNF, zinc finger protein.

on BTA1 which might explain this interaction. The term Marker Assisted Selection appears to have outlived its usefulness in that it emphasizes single markers whilst ignoring haplotypes and their interactions.

EBVs

From a background in population genetics, it was natural to develop indices of breeding values. There are now many versions of such measures as illustrated in textbooks such as Simm [24] who discusses the history and applications in some detail. In many cases, however, precise definitions are not given. What are the reliable advantages of EBVs v PBVs v PTAs v BLUPs v MACEs? What specific combinations of genes do these identify? Are they really surrogates for deep ancestry and therefore innumerable ancestral haplotypes? If so, how should the indices be used?

A timely review by Lester (personal communication 2013) emphasizes the continuing impact of R A Fisher's paper of 95 years ago. [4]

"The theoretical underpinnings of EBV estimation date back to 1918, when R.A. Fisher decomposed phenotypic trait variance into genetic and environmental components under strict assumptions of simple Mendelian inheritance and additive genetic effects across multiple loci. However, we now know that Fisher's assumptions do not hold for complex polygenic traits. For example, the abundant QTL literature clearly demonstrates that there are also a few loci (~ 4-6) with relatively large effects for any given, heritable trait. It is unfortunate that, while there has been considerable research and financial investment in EBV estimation, this has focused exclusively on data management/collation, and computation requirements for trait variance decomposition in extremely large, unbalanced datasets, rather than the development of more appropriate underlying genetic models. An additional concern is that EBVs, like heritability estimates, are poorly understood, and therefore may be misused. In statistical terms, EBVs are not directly observed, but are estimated, under strict assumptions of normality, as conditional random effects from a mixed effects regression model. In practical terms, this means that EBVs are simply a relative ranking of the sires within the dataset, and are very dependent on sufficient available progeny data, the other sires in the dataset, and the selection of the "correct" statistical model (in terms of both fixed and random effect variables) for estimation purposes. Therefore, EBVs do not represent a fixed estimate of the intrinsic genetic merit of a particular sire, yet this is how they are often interpreted, and used, in practice. Further, the "accuracy" measurements supplied with EBVs can also be misleading. Accuracy is explained to breeders as "the higher the accuracy value the lower the likelihood of change in the animal's EBV as more information is analysed for that animal or its relatives". While accuracy measures are a good indicator of the amount of progeny information for each sire, a sire's EBVs may in fact change, even given a high initial accuracy measurement, if (1) the statistical estimation model is changed in terms of the fitted fixed and random effects and/or (2) progeny data is added for new sires.

"An acknowledged problem with the use of progeny derived EBVs for sire selection is the necessary, and relatively long, lead time that is required to estimate the EBV for a given trait for a given sire. There are obvious advantages in being able to estimate genetic merit in young sires, and this, coupled with the release of the Illumina 50K bovine SNP chip, has led to the development of genomic (gEBV) prediction. The principle is that predictive equations relating SNP genotypes to progeny derived EBVs can be derived for a reference population with known EBVs, and that these predictive equations can subsequently be used to predict gEBVs from the SNP genotypes of individuals with unknown EBVs. This approach is deceptively simple, but not without some serious potential issues. Firstly, there is a statistical problem of identifiability with the development of the predictive equations as there are usually more variables (SNPs) than observations (animals). There are advanced statistical techniques to handle this problem (such as PQL, Lasso, Bayes B, Bayes C), and in spite of ongoing research, it is not yet clear which method is the most universally reliable, given that the underlying genetic architecture of the traits is generally unknown. However the most serious problems relate to the selection and size of the reference animal data sets necessary to build the predictive equations. Theoretical studies on the size of the reference datasets required for accurate gEBV prediction, suggest that upwards of 10,000 reference animals may be required [24]. Further, these predictive equations will be highly sensitive to any difference in the genetic background between the reference and test animals and therefore multiple large reference populations will be required. In this context, studies have already shown that, as expected, genomic prediction is more accurate when used for testing offspring of the reference animals compared to unrelated animals [24]. These issues regarding the size, and appropriateness of the selected reference populations may ultimately prove intractable.

"It is interesting to reflect that Fisher's 1918 approach to determining the genetic contribution to a trait, which required no understanding of biological and genetic mechanisms involved, still persists in this genomic era in relation to the use of EBVs and even gEBVs. The consideration of alternative approaches is long overdue."

Recently, Akdemir and Jannink (2015) have proposed an improvement to EBVs. They draw the distinction between "breeding" and "genetic" values. The former, as in EBVs, is restricted to additive effects of multiple genes assuming "free recombination". The latter includes epistatic or remote interactions and is therefore more in keeping with common sense expectations and with the largely neglected building-block hypothesis of evolution (Holland 1975).

Although apparently unknown to Akdemir and Jannick, it is now perfectly clear that the genome does contain polymorphic frozen blocks and that phenotypes depend upon interacting ancestral haplotypes. Any approach that allows for this reality is welcome, and especially so, when any grouping of markers is based on actual haplotypes as they segregate.

There remains the problem of demonstrating the benefit of one method versus another. Traditionally, the proponents of a value claim responsibility for the improvements in livestock over recent times. However, amazingly, the advocates have not controlled for other changes in selection and management and they carefully avoid reference to negative consequences such as decreases in fertility, as just one example.

Heritability

There is a vast library of attempts to determine heritability of phenotypic traits such as IQ. A prominent characteristic is continuing controversy.

The notion that "heritability" = Vg/(Vg + Ve) (where V is variance, g genetic and e environmental) has been adopted but often uncritically as explained above and as illustrated by attempts to calculate Vg and Ve.

Ayala [26] explains in his book entitled "The Big Questions: Evolution" that the twin method determines what fraction of the total variance is due to genetic differences. He asserts that all variance between "identical" (meaning monozygotic) twins is environmental, i.e. Ve, because they are genetically identical. Further, "fraternal" (meaning dizygotic) twins have half their genes in common so that their genetic variance is half that of the population i.e. the total variance is Vg/2 + Ve. It then follows that the difference between the variance of the two types of twins is Vg/2 + Ve - Ve = Vg/2. Hence double this difference, divide by the total phenotypic variance and you have "heritability" [26]!

The flaws relating to Ve are manifest, especially in assuming that Ve is the same in monozygotic and dizygotic twins and, remarkably, even within whichever population is used to determine total phenotypic variance. It is now well known that factors such as age, sex, hormones and nutrition have major effects on genetic penetrance effectively eliminating the use of Ve.

The notion of genetic variance is difficult to reconcile with current concepts. In general a large degree of scatter can be assumed to be due to complexity of the genetics or the environment and more likely both. Necessarily, a ratio of variances is difficult to interpret. More specifically, the twin method is seriously flawed. The idea that dizygotic twins, essentially siblings, have half the total genetic variance because they "share half their genes" is curious. Such sharing might be true in a Mendelian system such as a single diallelic locus with alleles 1 and 2 and for example 1,1 and 1,2 parents giving rise to 1,1: 1,2; 1,1 and 1,2 offspring but more generally one expects distributions reflecting heterozygous 1,2 parents and 1,1; 1,2; 2,2 siblings (or dizygotic twins) with percentages approaching 25, 50, 25. Half the siblings share 1 and 2 and therefore 100% but the other half share nothing. It depends on which genes and what is meant by genes.

In practice, at each of the polymorphic regions of the genome the children of *ab* and *cd* parents will be *ac*, *ad*, *bc*, or *bd*. Thus, if we had prior knowledge to indicate that only *a*, *b*, *c* or *d* are important, half the offspring will be positive and half negative. However, this approach ignores the genotype and therefore any positive or negative interactions in *trans*. In fact, all four genotypes are different. Consider the same at other regions which are independently segregating and it becomes obvious that dizygotic twins and siblings are vastly different individuals and most certainly not half of unrelated subjects.

It is surprising that so much attention has been paid to the estimation of "heritability" so defined. It is curious that some estimates have been applied commercially. Ayala states that "heritability" of milk production is 0.3

whereas meat production is 0.7. The latter ratio of 0.7 was achieved by controlling the environment and reducing Ve to the benefit of Vg. The former ratio of 0.3 might surprise dairy breeders who standardise the environment and pay handsomely for genetics. Presumably such a low heritability could simply mean that there is very little scatter, or low variance, amongst cows which have been essentially inbred over multiple generations. Or does it mean that the cows were so heterogeneous that some were more affected by the environment than others? Either way, it is difficult to see how the concepts underpinning "heritability" will assist in identifying genetics which will improve milk production.

To paraphrase Gillespie [18], as cited above, it must be time to test the tools of quantitative genetics "in a natural setting" and to determine pragmatically how advice to breeders can be more useful. At the same time, it is pressing to use the observations of breeders to test the validity of models of evolution and to respond to the challenge offered by Ball [27] on the 60th anniversary of the DNA helix which seems to have had very little impact on population genetics. There appear to be some blind spots in attempting to understand the workings of evolution.

Blind Spots in Applications of Evolution to Vertebrates

The *sine qua non* for evolution is inherited polymorphism. This is packaged as conserved ancestral haplotypes which must truly be the units of inheritance because they are transmitted *en bloc* with a complete set of noncoding sequences which control gene expression.

Mutations may be captured at speciation but tend to be lost or ignored if more recent.

Ancestral haplotypes are the alternative sequences found within polymorphic blocks which are separated by essentially monomorphic sequences. Meiotic recombination occurs between the blocks and therefore remixes conserved polymorphism at each generation. In this way, benefits accumulate and confer advantage and direction. The concept of anamnestic evolution contrasts with mainstream teaching which asserts that ongoing mutation is essential to create the variation which allows selection. To the contrary, there is sufficient conserved and pretested polymorphism to generate effective diversity and individuality.

Because there are multiple haplotypes at multiple blocks throughout the genome, there are at least trillions of diploid genotypes thereby ensuring that billions now alive - and their ancestors - are essentially unique [28]. Witness the difficulty in finding matched donors and recipients for transplantation and bone morrow grafting.

If, as we now know, relevant polymorphism is conserved in packages, it is distracting to focus on single nucleotide di-allelic differences largely outside the informative haplotype sequences which define individuals. This focus is driven by commercial returns.

Like Ball [27], we are all in awe of the complexity and incomprehensibility of the molecular processes which result in efficient inheritance and evolution. This sentiment, however, should be tempered by a recognition of the sheer beauty of the genomic system which creates diversity and memory at each generation. This is the power of the concept of Quantum Genomics whereas the dogma of selection of ongoing random mutation would lead to chaotic outcomes.

As a system, genomic evolution is truly brilliant in that it includes, *inter alia*:

- Redundancy by virtue of paralogous regions.

- Duplication as a method of increasing options and potency.

- Insertion and deletion as a means of reinventing combinations of sequences.

- Age and sex dependent penetrance.

- Retroviral integration associated with mutation at speciation.

All of these features have been known for decades and have been presented forcibly from time to time.

There must be explanations for blind spots becoming fashionable. One may be that the history is told in terms of personal discoveries and that the prospects for credit for the next great discovery may improve if competing ideas are ignored. Another explanation may be that groupthink and commercial returns are linked.

SNPs for GWAS

A further development from Population Genetics is the increasingly popular application of Genome Wide Association Studies using SNPs. The central idea is that a different nucleotide may directly or indirectly identify a gene or trait or disease.

Several studies illustrate the applications.

SNPs and Milk Production

Meredith et al. [29] studied two separate populations of Holstein-Freisians using 54,001 biallelic SNPs (Illumina Inc., San Diego, CA). However, 2,410 SNPs were removed because their position was unknown or because they were on the X chromosome. On the X, there was "large scale divergence between SNP genotypes and animal sex (i.e. numerous SNPs... were... heterozygous in male animals) indicating poor SNP calling". Another 230 were discarded because they did not conform to Mendelian inheritance. After editing out for various reasons (such as minor allele frequency less than 5%), there were 40,668 SNPs (75%) for study. These defined 10,958 "haplotype blocks" which averaged 97 Kb* in length and summed to approximately 1 Gb. Each "block" had an average of 2.84 SNPs giving a density of only about 1 per 30 Kb. Approximately 23% of the 40,668 SNPs were within approximately 2 Gbp between the blocks giving a density of 1 per 200 Kb.

* given as 97,006 Kb

The results supported previous findings and were consistent with the general impression that there are numerous QTLs for milk production. At least 6 chromosomes have been implicated including, BTA 14, 20 and 6, with potential candidates such as DGAT1, GHR, PRLR and Kappa Casein respectively.

After numerous corrections and adjustments to reduce potential errors, 276 "novel" SNPs were detected. These suggested a role for 45-49Mb of BTA 13 but no definite conclusions were possible at this stage. It is concluded that further studies are necessary. "In general, GWAS studies act as a screening tool, from which significantly associated regions can be further refined using higher marker density with the ultimate goal of identifying candidate genes".

Some of the queries which should arise are:

- Why are so many corrections and qualifications necessary?

- How useful are SNPs in identifying candidate regions?

- Why are so many regions identified?

- What is the distribution of SNPs throughout the genome?

- How will higher marker density help?

- What is the relationship between SNP derived haplotypes and those actually observed in segregation studies?

These queries will be addressed below after describing recent advances in genomics.

SNPs and Beef Production

Another SNP study is interesting for different reasons. Sellick et al. [30] show that there is a QTL for muscle at 0-15cM on BTA2 and propose that a base change resulting in F94L may be responsible for potential benefits in beef production. Expression varied between muscle groups. There were differences depending upon whether the F94L was carried by the maternal or paternal haplotype suggesting to the authors differing extents of LD. Although presented as an example where a single base change explains a QTL, with caveats, the ancestral haplotypes have not been presented and other linked polymorphisms could be expected. Undoubtedly, the usual complexities of regulation, penetrance and interaction apply to beef production.

Reproductive Biology

The revolution initiated by artificial insemination has changed livestock breeding in many ways. The sheer convenience of AI is very attractive to industry especially for example in dairy herds. The cows are monitored and available so they can be mated with preferred semen.

This convenience created the opportunity to upgrade but also a dependence on those promoting particular semen. Progressively the dairy industry has come to depend upon international suppliers at least partly because of the impression that common semen have meaningful indices and necessarily appear better than local alternatives. One potentially harmful effect is that the gene pool has converged leaving less opportunity for comparisons and for blending of whole genomes. Another consequence has been an industry wide concentration on certain measures which may yield immediate commercial gain without a longer term perspective.

These consequences have been further encouraged by embryo transfer and cloning. Now yet further opportunities are available through *in vitro* fertilization and embryo biopsy.

All of these developments are powerful and capable of delivering the desired outcomes. However, they depend critically upon the validity of the assessment of short and longer term benefits of particular semen. It may be that advances in reproductive biology have preceded an understanding of selection based on genomic inheritance.

Genomics

Surprisingly or not, the lessons of genomic analysis actually accord with the breeders' blending approach rather than with the teachings of population genetics or the practices of reproductive biology.

Here the term genomics is used to refer to the structure or architecture of the continuous sequence. Elsewhere, unfortunately and inappropriately, the same term has been used to refer to SNPs used in selecting of dairy bulls.

The essential features of the vertebrate genome can be summarised as follows:

- Meiotic recombination occurs between polymorphic blocks or quanta resulting in generational remixing *en bloc* rather than scrambling diversity.

- At each block there are numerous extensive ancestral haplotypes .

- The polymorphism is conserved from remote ancestors and provides precise memory (as opposed to what has been termed cryptic diversity).

- Duplication is an important component of ancestral haplotypes giving rise to gene families and interactions.

- Retroviral-like elements have been inserted and deleted and are critical in defining and distinguishing ancestral haplotypes.

- Through whole genome duplications there are multiple representations of many regions through out the genome. These paralogies contribute to redundancy.

- Ancestral haplotypes interact in *cis* and *trans* in addition to epistatically.

- Individuality is encoded rather than created by mutation.

These features can be summarised as "Quantum Genomics" so as to emphasise that conserved sequences are packaged for inheritance *en bloc* (Figure 2 to Figure 6 and Figure 8). Thus, we can see the attractiveness of blending the quanta rather than attempting to split the components into discrete genes.

On this background, it is possible to explain some of the queries which have arisen by application of population genetics, EBVs and GWAS using SNPs.

A useful approach is described by Smith et al [23] who specifically examined the distribution of nucleotide polymorphism over extensive sequences of ancestral haplotypes.

The density of SNPs varied substantially over the 4.5 Mb sequenced, reaching about 50 to 100 SNPs per Kb or 5 to 10% in the alpha, beta and delta blocks of the MHC.

Between these blocks, the frequency was less than 1% with some regions approaching zero. These extreme differences in density have been found in other studies [22, 31] and are undoubtedly correct. They contrast with estimates from commercial SNP-GWAS projects such as Meredith et al [29] who reported values such as 1/30,000 within and 1/200,000 between their LD "blocks". The difference undoubtedly reflects the completely different methods as explained below.

The MHC was selected by Smith et al. [23] because of prior knowledge which allowed segregating, independently verified, ancestral haplotypes to be characterised and compared. Nineteen examples of the 8.1 ancestral haplotype were almost identical. There were only 11 different SNPs over 3.6Mb being less than .0003% in contrast to up to 10% when different haplotypes were compared. This remarkable degree of conservation suggests, with certain assumptions such as a fixed mutation rate of 3 per genome per generation, that there was a "common ancestor" of 8.1 some 10-30,000 years ago. Other evidence is consistent with the idea that an ancestral population spread and flourished from an earlier time thereby implying that the mutation rate may be less than assumed.

The concept of a common ancestor, and especially the precise time of that ancestor, is traditional but somewhat quaint, at least when measured in thousands rather than millions of years. However, by all measures, it does appear that the diversity was established before the human expansion and that the 8.1 was just one of many. There must have been numerous ancestral haplotypes in order to explain the extreme polymorphism demonstrated by Smith et al [23] when different haplotypes were compared. Previously, we have shown that some, like 57.1, are found in all populations [32] in contrast to 8.1 which is restricted to Caucasoid populations [33]. Irrespective of the timing, all of these haplotypes have been conserved to a remarkable degree. Thus, there was a period during which the differences appeared followed by a period of conservation. For this and other reasons, we have proposed that retroviral infection played a major role in generating diversity and that integration of the viruses was associated with sustained conservation [22].

Even more interesting, however, in terms of mutation and recombination, is the explanation for differences within and between blocks.

Meiotic recombination occurs between rather than within polymorphic blocks. Ancestral haplotypes are conserved *en bloc* because meiosis is somehow inhibited by polymorphism. The consequence is that intact haplotypes are shuffled between the generations [34]. They are not scrambled or mutated and not reinvented as many have believed. All of the assumptions of Fisher and his disciples are invalid.

It may be that mutation rates appear patchy because there is inhibition of mutation within polymorphic blocks. By mutation, we mean a change which may be repaired or lost. If retained and captured within a frozen block, the change becomes an inherited polymorphism. Since the polymorphisms which define 8.1 and especially 57.1 are so conserved in all geographically scattered descendants of very remote ancestors, we conclude that there must be an active process whereby polymorphism is self conserving.

From these and many other studies of the human genome, it is clear that the commercial SNP approach, as used by Meredith, underestimates the amount and significance of polymorphism in the total genome. It is likely that most polymorphism detected by SNP typing is actually between informative blocks and therefore unlikely to be of direct value in relation to MAS or EBVs. Quantum genomics says that it is actually necessary to define the ancestral haplotypes by reference to sequences which have been observed to segregate over many generations and to determine how *cis*, *trans* and epistatic interactions affect expression of complex phenotypes. Contrary-wise, it is necessary to de-emphasize casual mutations which are of no causal evolutionary significance.

Sadly, to date, there is a dearth of information from all species other than Hosa. However, published, commercial and private data on cattle appear to be entirely consistent until proved otherwise.

The queries arising from Meredith's careful analysis, as listed above, may be addressed as follows:

Why are so many corrections and qualifications necessary to interpret SNP GWAS?

As illustrated nicely by the Smith study, it is critical to determine haplotypes (or phase) by segregation through generations. It is then possible to determine where recombination occurs, the length of the haplotype and its defining characteristics. Sequences can be aligned and compared meaningfully only after allowing for haplotype specific duplications, insertions and deletions. There are far more haplotypes that might be expected and no rational basis for considering some bases common or rare. Notions such as double hits (least frequent allele present at least twice thereby under representing rare SNPs) are not required in analysing genomic sequences. Errors should not be expected unless they arise from the analysis itself and are fundamentally conceptual rather than technical. Smith et al. [23] found only 25 SNPs which could identify the 8.1 haplotype alone but these account for less than 1% of all SNPs and could not be recognised without prior knowledge of the haplotype, again illustrating the enormity of the task for GWAS.

How useful are SNPs in identifying candidate regions?

There appears to be a degree of consistency in the studies of milk production suggesting that the numerous candidate regions have been identified. At issue is how to proceed. Are all regions interactive and equally important? Which regions are similar in some respect?

Why are so many regions identified?

The vertebrate genome has been subject to whole genome duplication and dispersal so that some sections are represented up to 4 times. These paralogous regions have evolved differently but still share similarities in function, disease association and genomic structure. It seems that redundancy has allowed divergence of function with, for example, the development of enzymic cascades and protein families. It can be expected that critical functions such as milk production will be controlled by haplotypes of at least several paralogous regions.

What is the distribution of SNPs throughout the genome?

Patchy, based mainly on human studies but also apparent in some GWAS in livestock. Nucleotide diversity is high in polymorphic frozen blocks and much lower elsewhere. It follows that informative polymorphism will be found in these blocks rather than in between. Fortunately, this implies that most benefit will be obtained by testing only a small proportion of the genome. Elsewhere, in McLure et al. [28], we recommend a focus on the most polymorphic 10%. Unfortunately the GWAS approach concentrates on the remaining 90% and greatly underestimates the informative content of segregating polymorphic frozen blocks.

How will higher marker density help?

The aim should be to identify ancestral haplotypes within polymorphic blocks. Higher density, without haplotype definition, will be more, not less, confusing and undoubtedly more expensive.

What is the relationship between SNP derived haplotypes and those actually observed in segregation studies?

SNP blocks identified by LD are confusing if the intention is to identify heritable segregating ancestral haplotypes i.e. conserved sequences. Since LD depends on associations in excess of those expected by chance, the results are affected by relative frequencies which can vary widely.

Consider two linked loci, A and B. At each there are gene frequencies of A1=0.9, A2=0.1, B1=0.9 and B2=0.1. Under the free recombination model the expected frequencies of the combinations will be:

A1, B1	0.81
A1, B2	0.09
A2, B1	0.09
A2, B2	0.01

The demonstration of significant departure from expectation (or linkage disequilibrium) will be affected by:

- The relative frequencies.
- The number of alleles and how these are shared between haplotypes.
- The number and type of errors, (see above).
- The number of loci.

These problems remain intransigent but, even more important, is the fact that LD does not necessarily identify common sequences and therefore ancestral haplotypes. To the contrary, as emphasised for many years [22,35] there are false positives and false negatives. Ultimately,

- segregation through multiple generations, and/or
- sequence identities between the marker loci

must be demonstrated before assigning ancestral haplotypes.

Once identified, it can be seen that particular ancestral haplotypes are defined by the *cis* combinations irrespective of the frequency of individual components. The misconceptions amongst quantitative geneticists are really surprising. The block and haplotype structure of the vertebrate genome was well understood two decades ago [35, 36] but must have been unknown to those who used LD to define different types of blocks and who attached different meanings to old terms like extended and ancestral haplotypes. See for example the cover, Fig 10.7 and p557 in Hartl and Clark 2007 [19].

An Alternative Strategy For Selective Breeding

So, to take advantage of the accumulated experience of breeders and quantitative geneticists whilst also exploiting advances in genomics, we recommend the following eleven steps towards a synergistic strategy:

1. **Define** the blocks and their ancestral haplotypes in regions of interest (Figure 17) including those identified by SNP GWAS. Dot plots should be included.

2. **Develop** robust haplospecific markers which allow at least more frequent haplotypes to be distinguished.

3. **Confirm** segregation in multi-generation families (Figure 18).

4. **Identify** those haplotypes which are relatively frequent in breeds of relevance to the function, trait or disease of interest whilst classifying those which are common to multiple breeds and unlikely to be critical in explaining differences between breeds (Figure 19).

5. **Confirm** breed markers by demonstrating high predictive values for breed identification.

6. **Develop** minimal requirements for reliable determination of breed, parentage and, consequently, traceability and food safety.

7. **Compare** frequencies in elite versus poor performers (and their forebears) allowing for vagaries of penetrance. It is critical to allow for the diploid or *trans* mix of haplotypes rather than considering only either the maternal or the paternal haplotype. Consider *cis* interactions.

8. **Compare** candidate haplotypes to determine whether these have arisen by ancestral recombination events which can help to localise the operative components.

9. **Design** AI, ET and Cloning experiments to test hypotheses including those involving *trans*, *cis* and epistatic interactions with variable degrees of penetrance. Ultimately the aim is to blend preferred haplotypes into already successful herds or breeds or, using Marker Assisted Selection, to increase the frequency of haplotypes already present in some of the breed.

10. **Consider** functional explanations for observed haplotype associations. Based on the human experience, complexity makes this a difficult task. Within the MHC, there are hundreds of strong associations but only a few have been explained. In each case the mechanism was largely known before sequence was available. Fortunately, benefits can be obtained without the need to understand the mechanism.

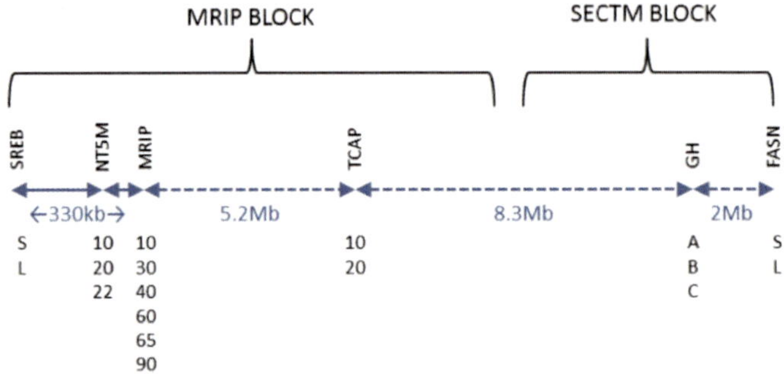

Figure 17
Relevant Genes, Markers and Blocks within a 16 Mb region on Bota Chr19.
The key PCR markers are shown with their polymorphic alleles viz SREB, NT5M, MRIP, TCAP, GH, and FASN. The two major blocks between which recombination is common are indicated, "MRIP" and "SECTM". Relevant genes involved in growth, muscle development and fatty acid metabolism in this region include: SREBF1, sterol regulator element binding transcription factor 1, involved in cholesterol and fatty acid biosynthesis and regulation intracellular lipids; STAT5A, transcription factor regulation of mammary tissue development, milk-fat percentage, fat cell formation and function; GH, growth hormone, breakdown of lipids; Urotensin2 receptor, involved in glucose metabolism, insulin resistance, skeletal muscle fat deposition, fatty acid metabolism; FASN, fatty acid synthase, involved in the complex enzyme system involved in fatty acid synthesis.

Figure 18
Confirm segregation in multi-generation families. MRIP and SECTM haplotypes as codes.
Williamson et al [38] and Lloyd S., Bayard D., Lester S., Williamson J.F. and Dawkins R.L. (2013). The Value of Haplotyping. InterBull meeting, Nantes, France, August 2013.

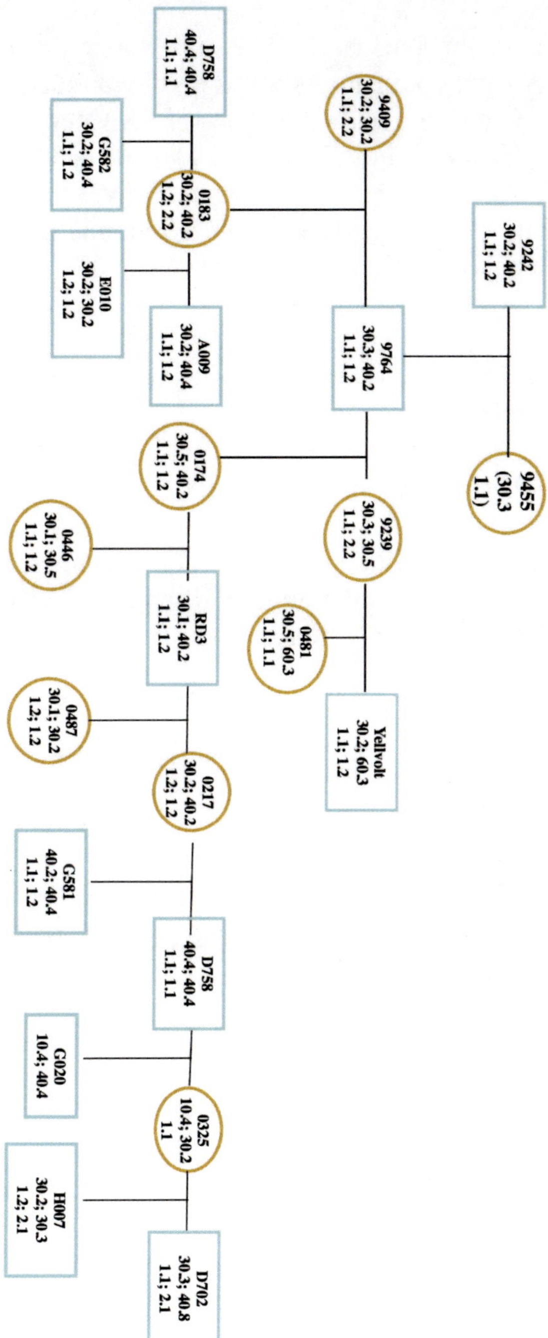
9455
(30.3
1.1)
9242
30.2; 40.2
1.1; 1.2
9409
30.2; 30.2
1.1; 2.2
9764
30.3; 40.2
1.1; 1.2
9239
30.3; 30.5
1.1; 2.2
Yellvolt
30.2; 60.3
1.1; 1.2
0481
30.5; 60.3
1.1; 1.1
D758
40.4; 40.4
1.1; 1.1
0183
30.2; 40.2
1.2; 2.2
A009
30.2; 40.4
1.1; 1.2
0174
30.5; 40.2
1.1; 1.2
RD3
30.1; 40.2
1.1; 1.2
0217
30.2; 40.2
1.2; 1.2
D758
40.4; 40.4
1.1; 1.1
0325
10.4; 30.2
1.1
D702
30.3; 40.8
1.1; 2.1
G582
30.2; 40.4
1.1; 1.2
E010
30.2; 30.2
1.2; 1.2
0446
30.1; 30.5
1.1; 1.2
0487
30.1; 30.2
1.2; 1.2
G581
40.2; 40.4
1.1; 1.2
G020
10.4; 40.4
H007
30.2; 30.3
1.2; 2.1

11. **Avoid confusion**. The importance of conceptual and semantic confusion is illustrated by a recent paper by Beatriz Cuyabano and colleagues of the Centre for Quantitative Genetics and Genomics, Department of Molecular Biology and Genetics, Aarhus University, Denmark [39]. The intention is to evaluate "Genomic prediction of genetic merit using LD-based haplotypes". The strategy is clearly stated:

- LD (with or without various manipulations) identifies haplotypes.
- Haplotypes identify mutations.

Of course, it all depends on what is meant by Linkage Disequilibrium, Haplotypes and Mutations. From the perspective of quantal genomics and adapting genetics, these two statements are incomprehensible and clearly indicate that the terms must have totally different meanings to Cuyabano et al and apparently to others applying quantitative genetics to dairy cattle.

The conceptual and semantic differences between the quantal and quantitative schools have prevented extrapolation from one to the other.

Quantal evolution

Here the emphasis is on haplotypes which have been observed to segregate in families, i.e. are faithfully inherited through multiple generations. The designation ancestral is applied when the haplotype has been conserved in remote descendants separated by hundreds of generations. The important implication is that recombination and mutation are suppressed often over many hundreds of kilobases. In humans it has been shown that these ancestral haplotypes regulate gene expression and function and are therefore critical to an understanding of what could be called genetic merit or quantitative trait loci. Indeed, many ancestral haplotypes have been discovered by studying the inheritance of disease and function. Linkage disequilibrium does not allow the recognition of such haplotypes because:

1. There may be tens or even hundreds of alternative haplotypes at a particular polymorphic frozen block.

2. Alleles at different loci within the blocks are often shared idiosyncratically by very different haplotypes.

3. Many defining alleles are extremely rare and therefore not in linkage disequilibrium with other common alleles shared by multiple haplotypes.

4. Retroviral elements, duplicons and indels are important in defining each haplotype.

As it happens, the polymorphism within these haplotypes is conserved and inherited. The term mutation refers to an observed or deduced change within a sequence. Necessarily most mutations occur between polymorphic blocks, are not components of ancestral haplotypes and of little relevance to complex functions regulated by ancestral haplotypes.

Renewal

In the absence of a renewal strategy, such as the above or better, breeders will remain sceptical of their advisors.

In Figure 19 we illustrate a simple way of beginning this renewal strategy. The principle is to define haplotypes (Figure 17) which yield a W plot (Figure 19) when two groups or breeds are compared. Those haplotypes centrally are shared between the two groups and include those which define the common evolutionary background. On either side are the haplotypes which are breed or group specific rather than shared. These are responsible for the distinctive features and the desired phenotypic traits undoubtedly in combination rather than alone. Such long haplotypes (>14Mb) are stably inherited in a Mendelian fashion through at least three generations (Figure 18). Markers which fail in this regard must be misleading.

The concept takes account of the prediction that robust heritable features will be due to specific haplotypes acting upon those shared by breeds and species.i.e. synergy between the new and the old.

The most useful regions will be those which permit the definition of haplotypes which yield distinct trimodal or W frequencies. For example, see the frequencies of TCAP-SECTM haplotypes on Bota chromosome 19 (Figure 17) between MRIP and FASN in Wagyu and Simmental Cattle. The distinctive W shape of this trimodal distribution shows that these haplotypes define both the common evolutionary background (shared haplotypes in the central peak) and the breed specific traits (associated with genotypes that include a haplotype in the outer peaks). Haplotype names reflect the occurrence in the breeds.

A comparison of the C19 haplotypes of Wagyu or Jersey and Akaushi show them to be entirely separate breeds, as different from each other as Wagyu are from Simmental (Figure 20).

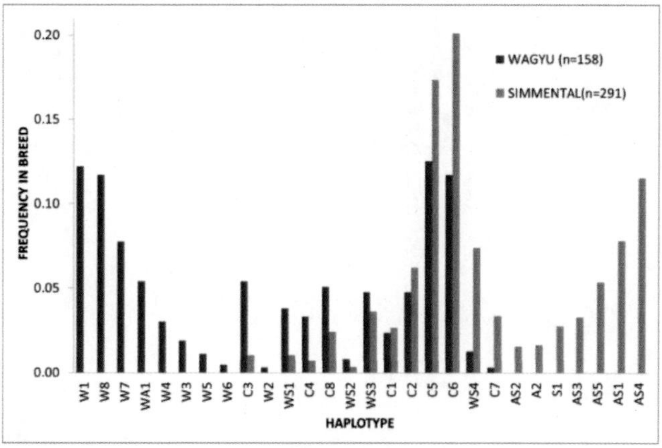

Figure 19
Shared haplotypes represent the common evolutionary background, exclusive haplotypes represent breed specific traits – the "W" plot.

As in Figure 18, MRIP and SECTM haplotypes are given as codes. See Williamson et al [38] and Lloyd S., Bayard D., Lester S., Williamson J.F., and Dawkins R.L. (2013). The Value of Haplotyping. InterBull meeting, Nantes, France, August 2013. (W=Wagyu; S=Simmental; C = present in all breeds; n=number of cattle).

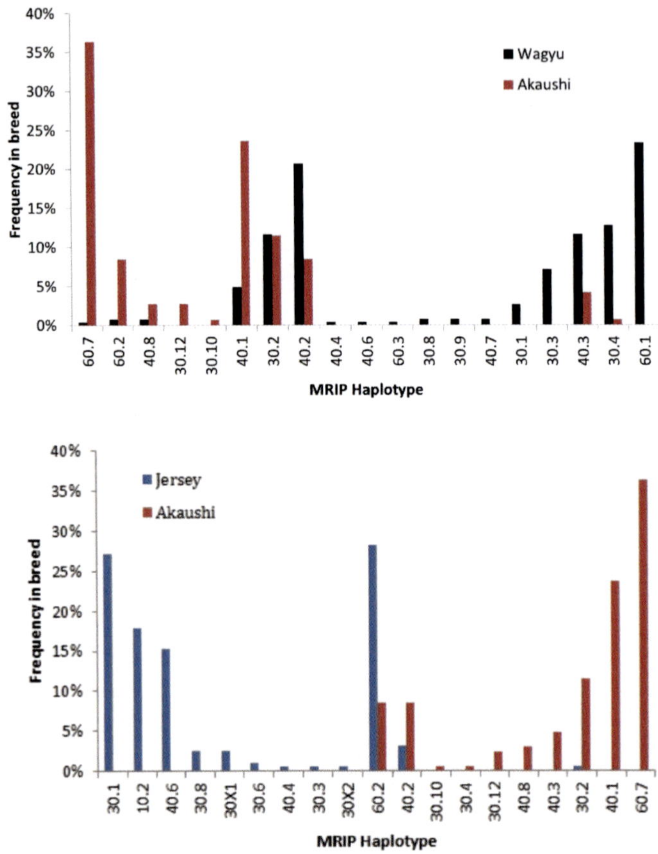

Figure 20
MRIP haplotypes as "W" Plots of Akaushi versus Wagyu (Black, top) and Jersey (bottom). Lloyd, SS, Steele, EJ, Dawkins, RL (C.Y. O'Connor ERADE Village Foundation Annual Report 2015)

Acknowledgments

These concepts were developed in association with the staff and students of Melaleuka Stud and C.Y. O'Connor ERADE Village Foundation. Special thanks to Sue Lester, Dom Bayard, David Blackmore, John and Keith Hammond, Scott de Bruin, David Bellamy, Alan Peggs, Sally Lloyd, Tony Lloyd, Charlie Stewart, Ted Steele, Peter Keating, Pat Carnegie and Craig McLure.

References

1. Lush, J. L. Animal Breeding Plans, Ed. 3. Iowa State University Press, Ames, IA. 1945.

2. Falconer, D. S. Introduction to Quantitative Genetics. Oliver & Boyd, London. 1960.

3. Hill, W.G. Applications of population genetics to animal breeding, from Wright, Fisher and Lush to genomic prediction. Genetics 2014 196(1): p. 1-16.

4. Fisher, R.A. The correlation between relatives on the supposition of Mendelian inheritance. Trans R Soc 1918 52: p. 399-433.

5. VanRaden, P.M. Efficient methods to compute genomic predictions. J Dairy Sci 2008 91 (11): p. 4414-23.

6. Goddard, M.E., B.J. Hayes, and T.H. Meuwissen. Genomic selection in livestock populations. Genet. Res. (Camb) 2010 92(5-6): p. 413-21.

7. Meuwissen, T. and M. Goddard. Accurate prediction of genetic values for complex traits by whole-genome resequencing. Genetics 2010 185 (2): p. 623-31.

8. Wray, N.R., et al. Pitfalls of predicting complex traits from SNPs. Nat. Rev. Genet. 2013 14 (7): 507-515.

9. Gianola, D. Priors in whole-genome regression: the bayesian alphabet returns. Genetics 2013 194(3): 573-596.

10. Fisher, R. *The Genetical Theory Of Natural Selection.* 1 ed. 1930: Oxford Clarendon Press. 272 pages.

11. Van Eenennaam, A.L. et al. Validation of commercial DNA tests for quantitative beef quality traits. J. Anim. Sci. 2007 85(4): 891-900.

12. Allais-Bonnet, A. et al. Novel insights into the bovine polled phenotype and horn ontogenesis in Bovidae. PLoS One 2013 8(5): e63512.

13. Mariasegaram, M. et al. Transcription profiling provides insights into gene pathways involved in horn and scurs development in cattle. BMC Genomics 2010 11: 370.

14. Capitan, A., et al. A newly described bovine type 2 scurs syndrome segregates with a frame-shift mutation in TWIST1. PLoS One 2011 6 (7): e22242.

15. Grobet, L. et al. Molecular definition of an allelic series of mutations disrupting the myostatin function and causing double-muscling in cattle. Mammalian Genome 1998 9: 210-213

16. Namikawa, K. Breeding history of Japanese beef cattle and preservation of genetic resources as economic farm animals, in Wagyu Registry Assoc. 1992: Kyoto, Japan.

17. Lloyd, S.S., Dawkins, S.T., and Dawkins, R.L. A novel method of measuring the melting point of animal fats. J. Animal Science 2014 J. Animal Science 2014 92 (10): 4775-4778.

18. Gillespie, J.H. Population genetics: a concise guide. 2nd ed. 2004, Baltimore, Md.: Johns Hopkins University Press. xiv, 214 pages.

19. Hartl, D.L. and A.G. Clark. Principles of population genetics. 4th ed. 2007, Sunderland, Mass.: Sinauer Associates. xv, 652 pages.

20. Dawkins, R.L., et al. Mutation versus polymorphism in evolution. Genomics 2013 101 (4): p. 211-212.

21. Khatkar, M.S., et al. Quantitative trait loci mapping in dairy cattle: review and meta-analysis. Genet. Sel. Evol. 2004 36 (2):163-90.

22. Dawkins, R., et al. Genomics of the major histocompatibility complex: haplotypes, duplication, retroviruses and disease. Immunol. Rev, 1999 167: 275-304.

23. Smith, W.P., et al. Toward understanding MHC disease associations: partial re-sequencing of 46 distinct HLA haplotypes. Genomics 2006 87 (5): 561-571.

24. Simm, G.. Genetic Improvement of Cattle and Sheep, Farming Press, Miller Freeman,Tonbridge, UK, 1998.

25. Goddard, M.E. and B.J. Hayes. Mapping genes for complex traits in domestic animals and their use in breeding programmes. Nat. Rev. Genet. 2009 10 (6): 381-391.

26. Ayala, F.J. The Big Questions: Evolution. The big questions, ed. S. Blackburn. 2012,London: Quercus Publishing Plc. 208 pages.

27. Ball, P. DNA: Celebrate the unknowns. Nature 2013 496 : 419-420.

28. McLure, C.A., et al. Genomic evolution and polymorphism: segmental duplications and haplotypes at 108 regions on 21 chromosomes. Genomics 2013 102 (1):15-26.

29. Meredith, B.K.,et al. Genome-wide associations for milk production and somatic cell score in Holstein-Friesian cattle in Ireland. BMC Genet. 2012 13: 21.

30. Sellick, G.S., et al. Effect of myostatin F94L on carcass yield in cattle. Anim. Genet. 2007 38 (5): 440-446.

31. Gaudieri, S., et al. SNP profile within the Human Major Histocompatibility Complex reveals an extreme and interrupted level of nucleotide diversity. Genome Res. 2000 10: 1579-1586.

32. Kay, P.H., et al. Coexistence of an MHC chromosomal segment marked by HLA B17, BfS, C4A6, B1, DR7, and DQw9 in different ethnic groups. Hum. Immunol. 1988 23: 27-35.

33. Degli-Esposti, M.A. et al., Ancestral haplotypes: conserved population MHC haplotypes. Hum. Immunol. 1992 34 (4): 242-252.

34. Gaudieri, S., et al. The major histocompatibility complex (MHC) contains conserved polymorphic genomic sequences that are shuffled by recombination to form ethnic-specific haplotypes. J. Mol. Evol. 1997 45 (1): 17-23.

35. Alper, C.A., et al. The haplotype structure of the human major histocompatibility complex. Hum. Immunol, 2006 67 (1-2): 73-84.

36. Alper, C.A., Z.L. Awdeh, and E.J. Yunis. Complotypes and extended haplotypes in laboratory medicine. Complement Inflamm. 1989 6 (1): 8-18.

37. Dawkins, R.L., P.H. Kay, E. Martin, and F.T. Christiansen. 1987. The genomic structure of ancestral haplotypes revealed by pulse field gel electrophoresis. In:

Immunobiology of HLA Vol I Histocompatibility testing 1987. Proceedings 10th International Histocompatibility Workshop and Conference. Dupont, B.(ed). Princeton, NJ and New York City, Nov 12-23 , 1987. Springer-Verlag. 1989 pp 893-895.

38. Williamson, J.F., et al. Genomic evolution in domestic cattle: ancestral haplotypes and healthy beef. Genomics 2011 97:304-312.

39. Cuyabano, B.C.D., et al. Genomic prediction of genetic merit using LD-based haplotypes in the Nordic Holstein population. BMC Genomics 2014 15: 1171 doi:10.1186/1471-2164-15-1171.

40. Akdemir, D and Jannink, J-L Locally Epistatic Genomic Relationship Matrices for Genomic Association and Prediction Genetics 2015 199 : 857–871.

****

Chapter 4

Individuality and Identification

*The shuffled uniqueness of genomes. The **why** and **how** of evolved individuality and its value to the individual and the species.*

- *Definition.*
- *Introduction.*
- *The model.*
- *The importance of cis and epistatic interactions.*
- *The history.*
- *An approach to the analysis of individuality.*
- *The importance of reproducible measurement.*
- *Significance of diversity.*
- *The Multifunctional Polymorphic Controller (MPC).*
- *Examples of individuality.*
- *Further implications of the Multifunctional Polymorphic Controller.*
- *Towards the pragmatic use of individuality for identification.*
- *Acknowledgments and References.*

Definition

Individuality is the aggregate of those features which are different when one individual is compared to another. Examples include face, voice, fingerprints and DNA to list only a few of many such features.

Introduction

The individuality of everyone is a fundamental belief in search of supporting evidence and unifying purpose. We are surrounded by examples of the depth of this belief;

- Each of us is inherently unique and uniquely special from birth.

- Diversity is necessarily good and creates healthy populations.

The persistence and strength of these and related convictions suggest that there must be persuasive evidence. However, credible data are hard to find and often qualified because of subjectivity and poor reproducibility.

Everyday experience teaches that individuals can be recognised and distinguished from others in many different ways and even in the most difficult circumstances, such as in a crowded airport. That is, it is as though everyone can ignore the overwhelming similarities whilst identifying unique differences.

Such effective recognition implies that each subject is not only unique but also somehow able to filter out uniqueness by comparison with others present and past. Here we ask, by analogy with electronics, what defining messages are sent and, when these are known, how these are received. Further, we hope to offer an explanation for the inheritance of send-receive processing over generations.

This facility is clearly shared with other mammals as illustrated by the uncanny ability of each cow to recognise its own calf and vice versa. We confirm the power of this ability following DNA testing of hundreds of cows and their calves reared in large mobs with many opportunities for confusion. Many calves are born on the same day. Most are indistinguishable to a human eye. Accuracy of the DNA match between mother and offspring approaches 100% with the exception of some twins. Frequently the mother will discard one of a dizygotic pair suggesting she can detect differences. Even when the mother and calf are unrelated, as in the case of embryo transfer, the recognition of each by the other is quite remarkable. Mother and calf must be individuals because they can recognise each other even though not genetically related. Thus recognition *cannot* be due to detection of *shared* polymorphisms or similarity as assumed previously. So, no matter ***why*** or ***how***, the cows and their calves must be distinct individuals able to recognise individuality.

The belief in the fundamental truth of individuality explains why even children are taught that they are special individuals and essentially, if not entirely, unique. It also explains the assumptions which underlie forensic investigation, legal and banking procedures and, more recently, the use of photo identification in airport security. The same acceptance of individuality underlies the manifest concern with "identity theft" and the popular culture of "finding the real me".

In spite of the strength of belief in individuality, there is still no explanation for the extraordinary diversity of so many features which contribute to identification and recognition. It is widely assumed that there must be unique characteristics generated specifically for each individual. However, there are many unresolved issues with the notion that uniqueness is created *de novo*. If so, how is it allocated? Is it possible to explain the sheer magnitude of the differences in, for example, voice, gait, face and fingerprints? Is it really true that everyone is an individual in all of these respects? And if so, ***why*** and ***how***?

The ***why*** may be approached by asserting that individuality is simply a *sine qua non* for the survival of mammals in the wild. Patently, each successive mother must be able to select and favour its own whilst rejecting others of similar age and appearance. Thus, in each generation, in turn, the mother must have been unique to the grandmother, seriatim. Differences sufficient to allow identification must be essential and must be maintained by processes which result in not the fittest, but *the most different*. If there were ongoing selection it would be *against similarity*. This would also favour heterogeneity in sibships, especially in multiple births and would encourage outcrossing in populations.

There are, of course many other potential advantages of diversity:

- Immune responses to pathogens.
- Survival in diverse environments.
- Friend-foe recognition.

These few examples add weight to popular beliefs in favour of heterogeneity although, interestingly, they do not explain the maintenance of differences to a degree which results in actual uniqueness.

Given that there may be plausible explanations for benefits of diversity if not individuality, the ***how*** remains an even greater challenge. As shown here, the current teaching of population genetics fails to provide an answer. Indeed, explanations for how individuality is created and maintained may well be the key to a new improved pragmatic understanding of evolution and genetics.

The Model

Uniqueness is encoded

Individuality is the direct consequence of inherited polymorphism. That is, there is such a pool of different DNA sequences within the population that every individual receives a unique genotype. Critically, at each generation, there is remixing of conserved polymorphic Ancestral

Haplotypes that were generated at speciation are inherited faithfully thereafter. Since there are hundreds of such sequences at hundreds of genomic locations (Polymorphic Frozen Blocks - PFB) and since the haplotypes at each block segregate independently, there are sufficient mixtures to explain the uniqueness of extant and preexisting individuals without the need to postulate errors of copying ("random mutations"). Because these blocks are widely distributed around the genome and because Ancestral Haplotypes regulate remote as well as local genes, the sequence polymorphism is translated into overlapping gradations of structure and function, which together account for all the quantitative differences which create individuality.

Uniqueness is advantageous to the individual as well as the population

The consequence of remixing conserved Ancestral Haplotypes is functional diversity. This permits complex biological pathways to perform adequately in different settings. There must be such an explanation for the certain fact that DNA encodes structures and functions over time and place and specifically as the environment changes and as the individual is challenged by growth, ageing and disease. In effect, there is adaptive memory which provides protection from past and therefore current challenges.

Uniqueness avoids the need for, and protects against, mutations

Importantly, the generator of diversity allows survival in the face of mutations in the less conserved germline between PFB. It also allows reserve capacity to deal with aberrations due to somatic mutations involved in cancer and ageing, for example.

An evolved strategy

By this model, individuality is a consequence of an evolved strategy for survival. In mammals, there are four components to the effector mechanism:

1. Generator of diversity

   At the time of speciation, new arrangements are generated largely as a consequence of retroviral elements. In this way, species specific polymorphic blocks are created.

2. Protector of diversity

   When a certain degree of polymorphism is established, intrablock meiotic recombination and mutation are suppressed thereby ensuring conservation of Ancestral Haplotypes and faithful transmission.

3. Supplier of diversity

   Through *cis* and *trans* mixtures of Ancestral Haplotypes, an extensive repertoire is available to generate graded and coordinated diversity of expression within and around the polymorphic frozen blocks. Meiotic recombination between the blocks is the main supplier through mixing a unique combination of sequences at each meiosis. This mechanism is enhanced by the complexity of the associated protein and RNA networks.

4. Distributor of diversity

   Transcription of coding and non-coding sequences within the PFB provides RNAs which distribute structural and functional diversity to create the individual phenotype.

Table 1 Lessons from Genomics

- Human diversity is inherited from ancestors rather than created by recent mutation.
- Diversity is regenerated at speciation and maintained by meiotic recombination between the ancestral haplotypes of nearby polymorphic frozen blocks.
- The unit of inheritance is the ancestral haplotype.
- Such sequences carry specific alleles, duplicons, indels and RLEs which together regulate gene expression.

Table 2 The Entrenched Dogma

- Genetic diversity or inherited variation requires ongoing mutation.
- Diversity accumulates through errors in copying of DNA.
- The unit of inheritance is the allele.
- At each locus alleles may be deleterious, beneficial or neutral.
- Meiotic recombination scrambles maternal and paternal alleles.
- Linkage disequilibrium between SNPs defines haplotypes.

These components of the model contrast with those underlying population genetics, Neo-Darwinism and molecular evolution as summarised in the tables above.

The thinking captured in Table 2 has resulted in some fundamental misconceptions. For example, it is often said that phenotypes depend upon genes shared according to the following ratios:

100% for identical or monozygotic twins.
50% for parent-child.
0% for unrelated.

Such thinking was based on simple Mendelian models as they apply to monogenic or, at most, oligogenic traits. For several reasons, analysing the heritability of complex phenotypes is another matter and extremely challenging unless the contributing haplotypes are used in segregation studies and ultimately characterised with all their interactions.

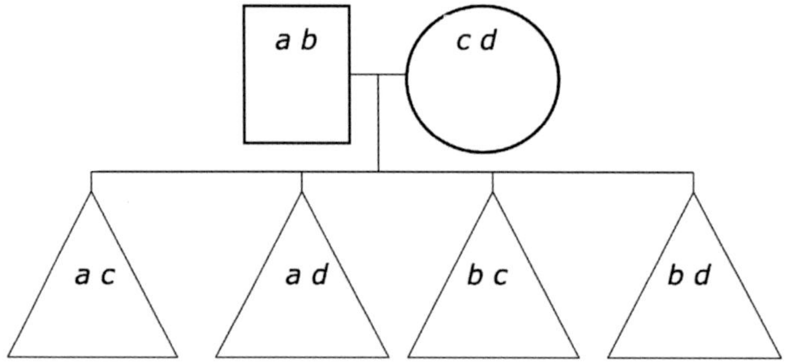

Figure 1
Segregation and Significance of Ancestral Haplotypes at a Single Polymorphic Frozen Block:
At any given PFB the father has two ancestral haplotypes designated *a* and *b*. The mother is shown as *cd*, assuming for the sake of demonstration that *a*, *b*, *c* and *d* are different Ancestral Haplotypes. The children must inherit one of the following genotypes *ac*, *ad*, *bc*, or *bd*. Note that the phenotype will depend upon the specific ***trans*** interactions between *a* and *c*, *a* and *d*, *b* and *c*, and *b* and *d*, such that additional diversity is created.

Firstly, although monozygotic twins have the same DNA, expression is dependent upon stochastic interactions of the many haplotypes at many PFBs and each interaction comes with redundancy and therefore the potential for differences in expression (see below). The more interactions, the greater the potential for difference. Complex phenotypes like IQ or personality would not be expected to be identical.

Secondly, 50% sharing of the total genome sequence by one parent and a child would only be expected if all four parental DNA sequences were absolutely unique but this is never the case. So, the degree of sharing is unpredictable. Frequently, the two parents will share some of many common Ancestral Haplotypes. Further, the *cis*, *trans* and epistatic interactions depend upon the precise complement of Ancestral Haplotypes at the different PFBs and this mix can ***never*** be the same because of the effectiveness of meiotic recombination in re-distributing all PFBs (see Figures 1 and 2).

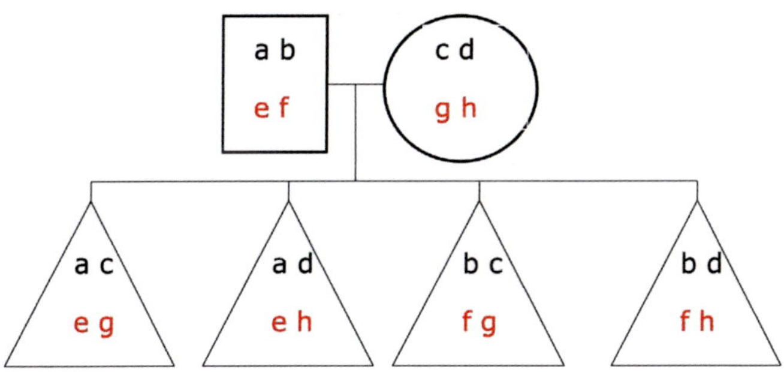

Figure 2
Segregation and Significance at more than one PFB:
The father has *e* and *f* in addition to *a* and *b*. Thus there are potential interactions between *a* and *e* , *a* and *f*, *b* and *e* and *b* and *f*. Similarly the mother has *g* and *h* in addition to *c* and *d*. Thus in the mother there are potential interactions between *c* and *g*, *c* and *h*, d and *g* and *d* and *h*. Note there are also four interactions in each of the offspring. Whether the AHs at the two different PFB segregate together or independently is a function of linkage being high if closely linked. Whether there are *trans* or epistatic interactions, there remain sixteen possibilities at each generation.

Thirdly, the sharing between siblings (and between dizygotic twins) may be 0 to 100% depending upon the PFB examined.

Fourthly, for the same reason, unrelateds can also be within the range 0 to 100%.

Consider as in Figure 1 a family where the father *ab* and the mother *cd* have four children *ac*, *ad*, *bc* and *bd*.

Although, on average, 50% of the siblings share *a*, *b*, *c* or *d*, any pair of genotypes may share 100% such as *ac* and *ac*. They may also share 0% such as *ac* and *bd*.

Further the phenotype will depend on the *trans* interactions between each haplotype so that a genotype such as *ac* may be completely different from *ad*. And these possibilities increase if we just consider the segregation and significance of Ancestral Haplotypes at more than one PFB (Figure 2). Given all 16 combinations and the potential complexity of all local and remote interactions (see Figure 3), it is highly misleading to teach that there will be 50% sharing of genotypes let alone phenotypes and that any departure from 50% excludes inheritance.

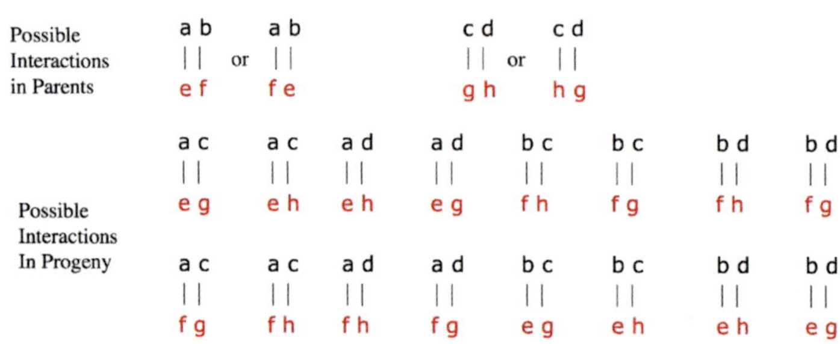

Figure 3
All possible interactions between PFB
(from Figures 1 and 2).

The Importance of *Cis* and Epistatic Interactions

Imagine that *a* and *b* are cogs within a gear box. Imagine also that *e* and *f* are meshing cogs within the same gear box. Thus, assuming heterozygosity, four gears are possible since *a* can mesh with *e* or *f*, and *b* with e or *f*. Some of the options are given in Figure 4. Meshed segregating geared loci are depicted schematically in Figure 5.

The importance of such meshing between interacting loci has not been appreciated in the past largely because the degree of polymorphism has been grossly underestimated. For example, as shown in McLure et al. [1], there can be up to a hundred Ancestral Haplotypes at each PFB. Assuming that each haplotype has different alleles at all loci within the haplotype, innumerable gearing ratios are available. Further, since there may be hundreds of *cis* interacting loci within each AH, there are yet further gearing ratios available. Yet further functional diversity arises through epistatic interactions, as shown by Lester et al [2].

Another implication of the gear box analogy is that homozygosity limits the diversity but augments the predictability of the interactions. This is why the use of inbred animals in medical research distorts the genetic reality. In an inbred strain, homozygosity may reduce scatter in experiments contrived to demonstrate the importance of single genes in isolation. However that approach pays no regard to the polymorphic implications of heterozygosity in the wild.

The gear box analogy provides an explanation for the phenotypic difference of monozygotic twins. As illustrated elsewhere monogenic disorders are largely concordant with high penetrance in contrast to complex diseases involving many interacting loci. Even though the monozygotic twins may have identical sequences at all of these interacting loci, the particular gear mesh actually expressed depends upon stochastic selection to some extent.

A further implication of stochastic meshing of polymorphism is that there is less attraction in invoking random mutation or epigenetics. That is, some of the features which are less predictable need not be attributed to errors or ill-defined interactions with the environment. Rather interactions are between polymorphisms encoded

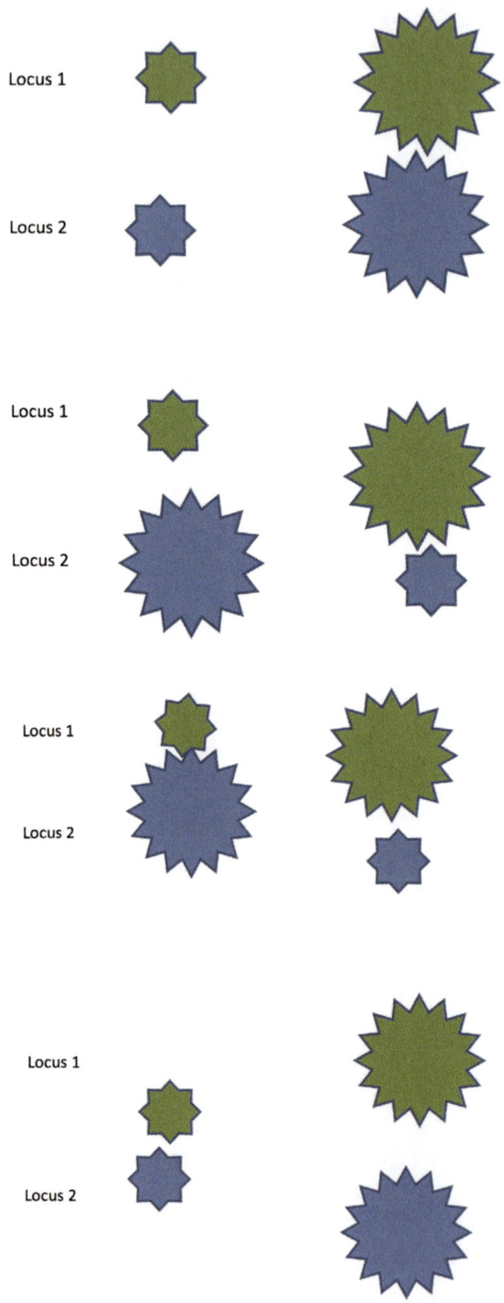

Figure 4
Two interacting gene loci depicted with large and small interacting cogs

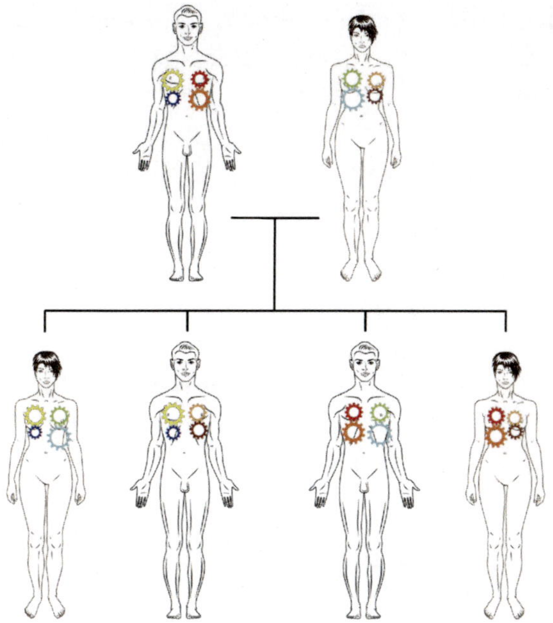

Figure 5
Segregating ancestral haplotypes depicted with the genes they carry interacting as meshing gears.
For comparison see Figures 4, 5 in Chapter 3.

within Ancestral Haplotypes and inherited. The range of options is determined by the range of genotypes within the population.

The History

Past explanations for individuality, from Pliny, the Elder, to the current practitioners of evolutionary psychology and behavioural genetics, illustrate the diverse way that beliefs and observations can become entrenched and even exploited for ulterior motives. Whether for the control of wealth, power or influence, including the divine right of kings, the concept of individuality has been accepted widely and imposed as necessary.

Here we focus on the last 100 years only and take just two examples which bear on the ***why*** and ***how*** but with contrasting conclusions. The basic facts have not changed. However, it is strange how fashions in evolutionary theory can dominate the interpretations and, as illustrated here, can result in neglect of the obvious. Worse, contrived arguments can lead to serious misconceptions and to political and commercial opportunism.

A worthy example is the book by Edward Lee Thorndike entitled 'Individuality' published in 1911 [3].

Thorndike reminds the reader that:

> "*A human being develops from, and is, an ovum... fused with a sperm... These (germs) vary amongst themselves*".
>
> "*The differences... are largely original, determined by the germs from which the individual develops... and... the ancestry from which he springs*".

Thorndike's emphasis on inheritance was largely based on his observation that certain differences are predictable, quantitative, usually continuous and amenable to statistical rules. Thus, in the first instance, such individual differences are predetermined and inherited from ancestors. It seems therefore that individuality is innate. For example, if average likeness is 0 and perfect similarity is 1, then father and son average 0.4, sibs 0.5 and twins 0.8. The proportions in relation to predictions from a Mendelian model, as discussed above, are not important but, there is a clear ranking order, implying some form of inheritance.

A contrast is provided by Olson [4]. The approach is based on finding mutations in the form of Single Nucleotide Polymorphisms (SNPs) using the Illumina technology. The conclusion is that genetic individuality is rare.

In the first case, Thorndike's very simple but powerful arguments fell upon a distraction generated by nature *versus* nurture and the American preoccupation with the concept of race. Thorndike showed that individuality was generated by inherited differences and therefore ancestry, but was attacked on the basis that inferiority was therefore predestined. These attacks conveniently ignored the unequivocal basic fact: all humans share a common deep ancestry. Of course, the depth of Thorndike's observations have become ever clearer, especially now that it is known that DNA sequences have been conserved from the time of human speciation.

In the second case [4], the misconception appears to have been driven by technology. The company is commercialising a kit approach which encourages the investigator to associate SNPs with recent mutations. Clearly, the approach ignores the fact that the relevant polymorphisms have been conserved, as envisaged by Thorndike 100 years earlier, and cannot be recent mutations.

These two case studies illustrate the difficulties encountered when basic facts and simple observations conflict with dominant beliefs and fashions of the period.

An Approach to the Analysis of Individuality

In order to consider the ***why*** and ***how*** of individuality, as distinct from ***whether***, it is appropriate to employ a systematic strategy which serves to distinguish between the two major alternatives:

1. Inheritance of differences through conserved DNA sequences.

2. Creation through *de novo* events such as recent mutations or errors of copying.

These two extremes can be contrasted as optimistic and pessimistic respectively. If polymorphism and individuality are transmitted faithfully and remixed at each generation, they can be refined, selected and utilised over time. If, however, random mutations and errors are involved, evolution is dependent on chance happenings and the descendants must be unpredictable and their future must be in doubt.

The problem remains that the pessimistic is implicit in the teaching of population genetics and molecular evolution. The more optimistic view appears to have been forgotten or discarded during debates around the edges. Adoption of concepts such as epigenetics can obscure the idea that individuality is inherited. Admittedly, until recently, there was a lack of a compelling explanation for how inheritance could be responsible.

Although frequently neglected, a very simple scheme was familiar to Thorndike and many others sometimes demeaned as biometricians and believers in eugenics. Inheritance can be assumed if a variable characteristic displays not a proportion but a *specific ranking* of the degree of *difference* from lower in identical twins through to higher in siblings and yet higher in unrelateds. By this logic the components of individuality are inherited. Even if nurture matters it must be acting upon deep ancestry.

"Identical" or monozygotic twins are often used as an argument for and against inheritance. In terms of transplantation outcome, monozygotic twins are identical and establish the importance of germ-line inheritance as opposed to recent mutation. At the other extreme, differences in the personalities of such twins have been used to argue that inheritance is not responsible for individuality. For example, J. R. Harris in *No Two Alike: Human Nature and Human Individuality* [5], stresses the differences in the personalities of identical twins and seeks explanations. After eliminating nurture and the like, it appears that the best explanation is that genetics was not well understood at the time!

However, whether considered identical or nearly so, depending upon the metric and the accuracy of measurement, it is obvious that the degree of inherited difference increases as these twins are compared with:

1 Dizygotic or “non-identical” twins,

2 Siblings and parents,

3. More distant relatives through to “unrelateds”.

The critical insight is that the differences increase in proportion to the number of meioses or generations of separation from the shared common ancestor. In monozygotic twins, there is no generational difference because a single egg divides after meiosis and fertilisation.

In dizygotic twins and siblings, there is a difference of one generation and therefore, on average, a separation time of 20 years.

For unrelateds and effectively all humans, the difference is thousands of generations and at least 20,000 to perhaps 100,000 years. For comparison, the chimpanzee is separated by millions of years and many hundreds of thousands of generations.

Surprising to many, individuality, however measured, is a compound feature which follows this ranking and is therefore at least largely inherited. Differences due to conserved polymorphisms must be accumulated and remixed rather than generated at each generation. This conclusion was not obvious before it was appreciated that diversity is created at speciation and then packaged and protected (see below and earlier Chapters)

It could be argued that *de novo* events such as mutations could accumulate progressively and achieve generational increases. However, if retained and if expressed in each generation, such mutations must have been converted to conserved heritable polymorphisms and been added to the ancestral pool for remixing [6]. It remains to be shown that this conversion does actually occur after speciation. It is possible that capture and freezing remain available to some degree, necessarily a relatively small degree, when compared to the mass retention of polymorphism generated at speciation and to the mass extinction of more recent mutations. In any event, evolution clearly depends on conservation rather than *de novo* mutations. This distinction is not semantic; it underlies any rational explanation for individuality as a critical component of mammalian evolution.

The concept of remixing at each generation has been emphasised from time to time depending upon the species of interest. Meiotic recombination after fertilisation has been central to any understanding of the evolution of mammals but usually in the sense that the end result is random scrambling and increasing the pool of differences. Here the emphasis is quite different. The effect of meiotic recombination is to remix conserved polymorphisms maintained within the ancestral pool. Diversity does not require scrambling or other *de novo* events such as mutation. Of course, there are other epigenetic factors such as environment, age, hormonal variation concurrent disease and viral infection. These can contribute to the appearance of complex and incomplete inheritance due in large measure to the fact that the conserved sequences are shared stochastically thereby allowing attributes to appear and disappear in particular descendants within a pedigree.

It is important to distinguish the current use of generational separation from the alternative applications used in population genetics as described in Chapter 3.

The Importance of Reproducible Measurement

Quantifying the differences which underlie individuality has been difficult. Reproducible testing will become important for many reasons but particularly because precise measurements can then be applied practically. When the differences are large, distinctions between individuals will be simple and the procedures will be credible and perform robustly. Contrariwise, when the differences are minor and more difficult to measure, there will be errors and disappointments. For this reason, a *quantitative* approach with levels of confidence is much preferred to the *qualitative* procedures which are used to confirm or deny the identity of an individual as claimed. When a bank, a court or an airport seeks to obtain a yes or no answer to identification, it must accept that substantial error is unavoidable - especially when attempting to distinguish between closely relateds.

For example, a close relative of the guilty party was convicted on the basis of DNA testing at the time but further testing revealed that 100 close relatives had the same test result. The reasons for such failures are clear. The method was over promoted commercially and was not a reliable measure of degrees of difference. The results are often interpreted in terms of probabilities raising the prospects of conviction of false positives. The preferred approach of defining true negatives is not widely understood.

A useful approach is suggested by immunogenetics as used to match donors and recipients for transplantation and bone marrow grafting. This is based on defining closeness of “tissue types” at the Major Histocompatibility or HLA region of the genome. Over time it became clear that ancestral haplotypes had been faithfully transmitted from remote ancestors and that differences relate to the number of generations of separation. Matching these haplotypes achieves successful outcomes in bone marrow grafting. As expected, matches are most likely when the donor and recipient are closely related. Thus, in principle, the results accord with the schema based on generational separation.

Which biometric differences can be quantified readily? The literature is replete with interesting developments and exciting claims but, unfortunately for an attempt to study their inheritance, very few proponents have shown that the testing is quantitative and reproducible. As possible, it will be important to relate the measurement of individual difference to generational separation. At the same time, better methods will become available for practical applications in security systems.

Significance of Diversity

Ultimately we hope to achieve an understanding of the mechanism of the *generator of diversity* and an explanation for the *active maintenance of individuality* at each generation.

It follows that genetics should provide the tools and that an understanding of evolution should provide the framework for investigation.

The lessons from the current teaching of evolution are really confusing. It is taught that differences are due to ongoing changes in the DNA or so called "random mutations". As one authority, Ayala [7] asserts

> "*...the process of mutation introduces new variation ...every generation...*" *and* "*...60 new mutations ..not present in the parents... .added to 3- 6 million polymorphic nucleotides per individual...*" *and calculates that* "*the total human population ...(creates) ...420 billion new mutations per generatio ... and ... (has) ...21-42 billion polymorphic nucleotides among the human beings living in the world today*".

It seems to follow that over say 10,000 generations, some 600,000 mutations have accumulated and now represent 10 to 20 % of the polymorphic nucleotides per individual. It is amazing that any descendant can survive! Everyone is a "mutant". Presumably the mutations are lost or ignored as discussed in previous chapters.

Many other experts, such as Lewontin [8] and Olson [4], stress the importance of new mutations as the source of variation or difference.

Thus, it appears that even very recently, there is overwhelming acceptance that mutation creates individuality. However, this is clearly incorrect.

Consider the following:

- New random mutations would create chaos if they accumulate progressively.

- Accumulated inherited advantage would tend to be diluted and even lost.

- Speciation is followed by relative conservation rather than continuing mutation.

- How could mutations be classified as new if they come and go as random errors?

- How could mutations be used in identification?

- How could family likeness be mediated by new mutations?

- Speciation is followed by relative conservation rather than continuing mutation.

- Why are matched bone marrow grafts successful?

The Multifunctional Polymorphic Controller

The concept of a multifunctional polymorphic controller (MPC) developed from the realisation that the polymorphic sequences which create diversity are inherited faithfully from generation to generation but are remixed to achieve individuality [1]. The intra-species ancestral pool available remains the same but the difference between individuals increases progressively as each descendant receives a unique subset.

The sequence polymorphisms are created at speciation, packaged and conserved as ancestral haplotypes at specific genomic regions or quanta known as PFB, and then distributed and mixed by meiotic recombination.

These Ancestral Haplotypes contain coding and noncoding sequences which are therefore capable of gene regulation as well as expression as polymorphic proteins. Through *cis*, *trans* and epistatic [2] interactions, functional as well as structural diversity is ensured. The consequence is that all individuals, other than monozygotic twins, are not only unique but unique in many respects. This concept can explain how conserved polymorphisms can directly control the very many aspects of individuality including those which are due to quantitative differences in expression.

The value of diversity may not be individuality itself but the potential for graded responses to challenges. In this way, opportunities are provided to members of the population or species. Whatever the selective forces, there must be advantages in having more options available without having to depend upon unpredictable random mutations. Rather, the changes required for survival have been accumulated mutatis mutandis during the successive speciations and then stored as memory.

Importantly, the diversity is comprehensive and assembled progressively so that functional consequences may adjust in a quantitative way and can be contributed to future generations. Effectively, the diversity of options provided by different ancestors provide enhanced memory for past challenges as they reappear over time.

Paradoxically, to some, because of the inherited reserve available to each successive generation, the species becomes more diverse in its options and also more relevant to its environment as it diverges over generations. This form of adaptation is powerful because the elements are pre-formed and pre-tested. The extent of the ability to survive and flourish is enhanced by the redundancy due to local and remote duplication, including paralogy. Through duplication and redundancy, the receptors become more promiscuous.

Table 3 Lessons from the MHC

- Age and sex related penetrance implying gene regulation of multiple pathways including many of no relevance to immunology.
- Susceptibility to disease is carried by extensive ancestral haplotypes; a few monogenic diseases are exceptions.
- Ancestral haplotypes determine clinical subsets, associated features and rates of progression.
- Important candidate sequences include indels, retro-elements, and variation in copy number rather than polymorphism within coding regions.
- Genetic redundancy is characteristic and permits advantages.
- Unlinked paralogous regions may contribute to complex functions and survival.
- Polymorphic control of expression, penetrance and regulation will be important clinically and biologically.

The concept of the polymorphic controller posits that genomic diversity, although encoded in quanta together summing to only a fraction of the genome, is actually reflected in essentially all functions. When first contemplated more than 20 years ago, the evidence was indirect and, not surprisingly, the idea has not received much attention. There has been a distinct preference to seek explanations for diversity of structure and function in terms of coding regions and therefore protein products. In the case of the MHC, it has been taught that the explanations involve presentation, recognition and processing of Class I and II molecules. Over time, however, the evidence for alternative explanations has increased.

By 1999, it was abundantly clear [9] that the influence of MHC was far greater than T cell recognition leading us to summarise the situation in terms of Table 3.

Since 1999, further evidence has accumulated and has been subjected to vigorous review. Only some of the less well known and largely neglected studies are summarised here. One feature in common to these conclusions is that they only really became possible after recognising ancestral haplotypes rather than HLA alleles, as the informative unit of inheritance.

In addition to its well known role with respect to autoimmune disease, the 8.1 haplotype affects such disparate functions as thymic hyperplasia, colonic cancer, IgA deficiency, mesangial nephropathy and ageing. The conclusion is inescapable; conserved polymorphism controls diverse functions which are not dependent upon HLA proteins.

- The 8.1 and thymic hyperplasia [10]. It has been known for several decades that this ancestral haplotype determines whether patients with myasthenia gravis have thymic hyperplasia rather than a cancer of the thymus.

- The 8.1 and cancer [11]. Toth et al. review conflicting results but show clearly that susceptibility to colorectal cancer is carried by the haplotype rather any single component yet identified.

- The 8.1 and ageing [12,13]. The Caruso group have continued their careful examination of influence of the MHC on non-immunological characteristics including longevity.

The mechanisms underlying these apparently remote non-immunological effects are now under active scrutiny. For example, the Knight group [14] have claimed to show that there are haplospecific differences in gene expression and, in particular, a zinc finger protein gene, ZFP57. Surprisingly, this paper ignores the earlier evidence for PFB and for the influence of the MHC on regulation.

Since it is now clear that similar insights may well apply to the many other PFBs discovered by Craig McLure [1], it is a small step to predict that the generalisation applies more generally to aspects of structure and function. Specifically, this may be the way to understand individuality. Conserved sequences at these diverse PFB may therefore relate to all biometric differences. AHs in various combinations may influence all pathways to some degree and could explain why there is so much natural variation, including extreme examples which cannot be attributed to polymorphism in the corresponding coding regions. The extent of these effects are yet to be determined but we predict that PFBs can explain much of the complex biometrics found in such diverse examples as facial features, fingerprints, iris patterns, hair colour to list a few.

The effects of the polymorphic controller must be maintained actively as the various polymorphisms are controlled during cell and tissue turnover for example. Eventually, however, there may be gradual failure in association with ageing. For example, innate differences in facial features and hair colour eventually decrease.

Because innumerable functions can be regulated and expressed quantitatively, the implications of individuality are yet to be appreciated fully. However the capacity to recognise the differences between individuals is of immediate interest. Multiple structural and functional differences create the potential for polymorphic recognition of the unique features. By analogy, with a receptor and its ligand, both are polymorphic thereby increasing the number of interactions. Therefore, it may not be surprising that the MHC region contains polymorphic olfactory receptors, hormones, mediators, complement components, etc. Thus PFBs may be responsible for the uniqueness of the individual and also for the capacity to recognise and distinguish that individuality.

Indeed in the context of "blending inheritance" Charles Darwin in 1857 said to Thomas Huxley:

"*....a sort of mixture... of innumerable individuality, as each parent has its parents and ancestors:- I can understand no other... way... crossed forms go back to ancestral forms*" (www.darwinproject.ac.uk/entry-2166).

Opportunities to test some of these implications arise through transplantation. If the donor and recipient are close genetically, the transplant will tend to be successful. So too would the master controller be effective. When the donor and recipient are more disparate, the transplant tends to be rejected and the master controller will be less effective because of incompatibilities. Rejection is due to a disturbance of regulatory homeostasis as well as immunologically mediated processes.

Another possible example is provided by Graft v Host disease after bone marrow grafting. There are so many aspects to the syndrome including some which suggest a change in gene regulation. Fibrosis is just one example.

Inspired by his observations of transplantation, Medawar recognised the importance of individuality in evolution. For example, in his monograph, "The Uniqueness of the Individual" first published in 1957, Medawar [15] says:

"*Although the inborn differences between human beings are combinational in origin... yet the combinants are so numerous, and so generous are the ways in which they may be combined, that every human being is genetically unique.*

" *Inborn diversity makes for versatility in evolution... not only for the present but also for what may happen...*

only those... survive... which, in the past, were versatile enough..."

In his 1970 book "Immunological Surveillance" MacFarlane Burnet [16] discussed the issue as follows :

"*Instead of regarding the lymphocyte as essentially an immunocyte concerned with the recognition and eventual elimination of foreign material.,. Burwell's (1963) contention (is) that the primary function of the lymphocyte is to mediate control of growth and morphogenesis. This is the function of '"mitotic control proteins" (MCP) which may be carried on the lymphocyte as a surface component or liberated into the circulation probably in the form of an α 2 globulin. Every distinctive tissue in the body... carries "tissue coding factors" [TCF) on cell surfaces... identical with and therefore "recognizable" by the corresponding MCP*" [17].

One comment on this speculation would be that the mobile white blood cells of the haematopioetic system are ideal physical distributors of the Multifunctional Polymorphic Controller so that polymorphism is available when and where required. The MHC was barely known in 1970 let alone 1963. Over subsequent decades attention focussed on transplantation and immune responses with scant attention to Burnet's proposal. The emphasis on the mobility of lymphocytes may have been distracting because the control may be mediated by the genotype rather than whether or not the cell expresses particular surface proteins.

Examples of Individuality

Facial recognition

"*The human features.... are so fashioned that... there are no two in existence who cannot be distinguished from one another*" in Pliny the Elder, Natural History, book VII quoted by J. R. Harris [3].

The ability to recognise an individual by facial features implies that there is something unique to see and

Figure 6
Facial Recognition

that the brain is able to compare what is seen with some form of pre existing library (Figure 6). This figure is displayed in trains with the intention of indicating that the transport officer is similar to the policeman. The unintended effect (on the public under camera surveillance) is that facial recognition is unreliable. It is certainly subject to many variables [18].

Fingerprints

Patterns of fingerprints are used widely and often decisively in forensics. The patterns are truly complex and difficult to record. Some features like whorls, loops and arches can be catalogued but there are numerous variants and degrees. Less pronounced features or minutiae include ridge bifurcations and endings.

For more than 200 years it has been shown that the precise pattern is retained and unique to every individual. The same appears to be true of other primates.

Given the complexity, there are diverse systems of recording and comparison. Reproducibility varies in different jurisdictions and has become a sensitive issue. As in other areas of biometrics, the more difficult to measure, the more the judgement tends to be attributed to subjective experience of experts.

In spite of these difficulties, there is no doubt that the pattern is inherited and the degree of difference follows the general rules of generational separation and accumulation of difference. Identical twins have essentially the same patterns. Epigenetic differences may be superimposed. Changes in minutiae occur with age, disease and trauma as expected. As shown by Cummins and Midlo [19] the differences are greatest in individuals from different races, through unrelateds of the same race, through parent child comparisons through siblings to maximal similarity in monozygotic twins. See also Jain, Prabhakar and Pankani, Biometrics.cse.edu.au [20] and Bodmer and McKie [21].

Confirmation of the influence of inheritance on fingerprints is provided by their absence in familial adermatoglymphia, [22] now attributed to deficiency of the gene SMARCAD1 mapping to 4q22-23 and within the 5.1Mb interval between microsatellites markers D4S423 and D4S1560 [23]. See also the ectodermal dysplasia inherited variants, Naegeli-Franceschetti-Jadassohn and dermatopathia pigmentosa reticularis. Clues to mechanism may be provided by the inhibitory action of Capecitabine [24].

How could such complex yet individual patterns be inherited? Combinations of two possibilities might be considered:

1. Direct expression of polymorphism.

By comparison with expressed polymorphisms, such as HLA proteins, it could be expected that there might be a genomic sequence for every fingerprint pattern. Whilst this possibility appears unlikely, it is noteworthy that combinations of HLA proteins do create polymorphism equivalent to fingerprints in that all unrelateds are essentially unique. Their proteins are encoded within only several Mb at 6p21.3 so it is possible that there is an equivalent fingerprint region still to be discovered.

2. Indirect variation through polymorphic regulation.

By comparison with more complex phenotypic variation, it is possible that the fingerprint pattern is created and maintained by the proposed generator of diversity. By this concept, the differences are due to degrees of regulation of epidermal structure.

Distinction between these possibilities might be approached through examining fingerprints after bone marrow and hand transplantation. In the later case the pattern would be expected to change rather rapidly. Szaherka et al. [25] showed some differences in the minutiae of a transplanted hand within months.

Odour

In their essay "Human Body Odour Individuality" Lenochova and Havilack [26] adopt a pragmatic approach. The evidence can be summarized as follows

1. Sex differences

Males and females can be distinguished to a greater degree than individuals. Males give an odour which is "musky, strong, and unpleasant". By contrast females are "sweet and pleasant". However, to compound matters, females are more pleasant during the fertile component of the menstrual cycle.

2. Individual differences.

The results have been disappointing although some suggest that individuals can sometimes be distinguished under certain conditions.

3. Kin recognition.

The results are controversial but there is some evidence for inheritance and/or learning. Attempts to demonstrate similarities within close relatives appear to have suffered from the lack of segregation studies (Figures 2, 3).

4. Mother-infant recognition.

The available information suggests that mutual learning plays a big role and may complicate the analysis.

The inherent complexity of odour and its recognition appears to have been largely ignored by studies which are based on a simple receptor-ligand model. There is some evidence that the MHC may be involved, at least in mice, but current information fails to explain changes during menstrual cycle and mother-infant recognition. It seems most likely that the effect of the MHC and other polymorphic frozen blocks are mediated through the Multifunctional Polymorphic Controller. Thus, in keeping with the lessons from the MHC genomics (Table 3) it would be possible to explain:

- Male *versus* female differences.
- Changes during growth and development.
- Hormonal effects.

- Low penetrance.

- Species specific differences between vertebrates.

Further Implications of the Multifunctional Polymorphic Controller

Protection from inbreeding and recessive disorders

The quantal structure of the genome ensures that each meiosis renews heterozygosity by remixing conserved haplotypes at all polymorphic blocks. Since there are multiple haplotypes at multiple blocks, the combinatorial options are such as to create structural and functional polymorphism even between siblings, dizygotic twins and even between monozygotic twins (see Figure 3). The MPC distributes this diversity and imposes options within biological pathways. Since there is polymorphism of heterozygous reactants, best illustrated by protein cascades, the consequence is a certain robustness of responses and therefore a reduced dependence on any particular gene or haplotype. The family likeness between members of a nuclear family reflects the partial sharing of conserved ancestral haplotypes (as shown in Figs 2,3) and survives without abrogating the benefits of the MPC and individuality. In this way there may a degree of protection from the expression of recessive disorders.

Protection from somatic and germline mutation

The quantal structure of the genome and the remote effects of MPC provide protection from dependence on single genes and haplotypes even when these arise by somatic or germline mutation of sequence outside PFBs. Additional protection is available through the redundancy created by segmental and paralogous duplication as perfected in vertebrates. Since each individual is unique, the degree of protection, and therefore penetrance, will vary stochastically. Thus, it may appear that the consequences are due to random mutations and environment factors. However, an alternative explanation is that the MPC creates the diverse outcomes without requiring mutation.

Survival of complex genomes over generations

The certain fact that relatively simple genomes have survived over millions of years and even more generations confirms the power of the processes which conserve DNA sequences even when exposed to extreme changes of climate. How this is achieved remains to be explained. More complex genomes have also survived over fewer but countless generations in spite of such manifest potential for error and exploitation by viruses and other parasites. In this context, it is not so much the conservation of sequence but the maintenance of complex interactive functions which requires attention. How have 8 billion individualised humans and their ancestors managed over even thousands of generations?

The discovery of the quantal structure of the genome provides an approach to this question. The conservation of the sequences of Ancestral Haplotypes becomes less remarkable than the explanation for the means by which these sequences interact to achieve individuality, on the one hand, and robust survival on the other. Conserved sequences must themselves permit adaptation to changing circumstances. One factor must be the re-mixing or re-blending of ancestral haplotypes by non-random meiotic recombination. The concept of the MPC adds another possible explanation. The interactions between heterozygous polymorphisms result in greater capacity and graded responses. Thus, according to the gear box analogy (Figures 4, 5: Meshing Cogs), each generation has an extra repertoire which applies to the regulation of pathways controlled by *cis*, *trans* and epistatic interactions (Figure 7: Interacting blocks and see Figure 3).

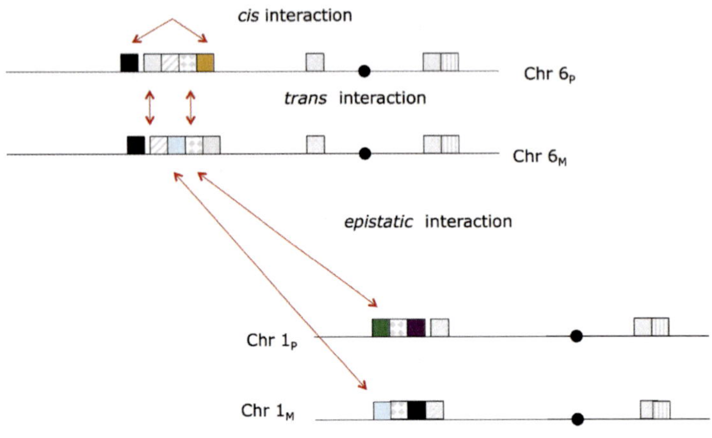

Figure 7
Interactions between frozen ancestral gene blocks of different chromosomal loci.

Optimal population size

One implication of the MPC concept is that there may be a limit to the amount of diversity. Up to that limit, the extra capacity is beneficial in providing adaptation without threat to the necessary conservation of sequence. Beyond that limit, the stochastic interactions may become excessive. The size of the interbreeding population is one determinant of the degree of permissible polymorphism. If the number of Ancestral Haplotypes increased indefinitely, the interactions would be experimental rather than pretested. Thus interbreeding may become less efficient as, for example, between horses and donkeys.

Bilateral discordance

Are there indicators of the limit permitted within the MPC? It would be of some interest if it were possible to recognise clues to excessive diversity. One possibility is bilateral discordance. For example, as shown in Chapter 3, readily detectable phenotypes such as polling in cattle and sheep, become less predictable in crossbreds and, it appears, subject to bilateral discordance. Note the illustrations of a horn on one side but not the other. Note also how the unilateral horn appears dysregulated.

According to the MFC concept, this anomaly could be a simple consequence of overly complex interactions between polymorphic products, such as the various clusters of keratin genes and their regulators. No doubt there are other explanations.

In a study of age-related cataracts, Hammond et al [27] found that differences between right and left are similar to those between twins whether mono or dizygotic. The range of difference between right and left provides an estimate of bilateral discordance attributable to stochastic variation within the MPC.

Speciation as a response to excessive diversity

The MFC concept implies that there are limits to diversity after which interbreeding becomes less efficient.

Each population is defined by the polymorphism created during formative "speciation" events. Similar populations may arise sequentially as appears to be the case in humans and cattle [28]. Interbreeding between these leads to sharing of conserved ancestral haplotypes. Eventually, however, the efficiency of interbreeding declines until there is a "speciation" event.

Towards the Pragmatic Use of Individuality for Identification

Although the benefits of individuality are plausible the precise benefit of each component remains moot. Clearly there is some value in the recognition of face, voice and odour, but the send-receive processes remain poorly understood and hardly amenable to use by police, banks and airport security. The techniques used by default, such as iris and fingerprints, appear to be of no biological significance and therefore difficult to apply in any comprehensive fashion. Other possibilities such as comparisons of DNA are promising but still require a theoretical framework based on an understanding of current genomics. For example, direct interrogation of Ancestral Haplotypes, as in transplant matching, is preferable to current approaches based on SNPs and microsatellites.

Sadly it appears that there are two approaches to identification:

- Firstly, those of biological importance but these are too complicated for the authorities in charge of forensics and security.

- Secondly, those of forensic importance but no biological significance, and therefore of little interest to the research community.

The Multifunctional Polymorphic Controller has the advantage that it can control and regulate innumerable phenotypic differences. Thus the most promising approach to identification of the individual is to detect the "Controller" itself. Fortunately the methods already exist and can be applied with access to only trivial amounts of DNA; as shown by Laird et al [29] there is sufficient DNA within fingerprints to allow haplotyping. A small library of reagents, as published by McLure et al [1], will permit identification with a degree of confidence in excess of that provided by conventional methods. Indeed by increasing the number of assays the confidence approaches 1 in 8 billion.

Acknowledgements

I am grateful to Craig McLure, Sue Lester, Louise Smyth, Sally Lloyd, Ted Steele and Sam Dawkins for their interest and advice.

References

1. McLure, C.A., et al. Genomic evolution and polymorphism: segmental duplications and haplotypes at 108 regions on 21 chromosomes. Genomics 2013 102 :15-26.

2. Lester, S et al. Epistasis between the MHC and the RCAa block iin primary Sjogren syndrome. Ann. Rheum. 2008 Dis. 67: 849-854.

3. Edward Lee Thorndike. Individuality. Houghton Mifflin, Boston 1911.

4. Olson, M.V. Human Genetic Individuality Ann Rev Genomics Hum. Genet. 2012 13: 1-27.

5. Harris, J.R. No two alike: Human Nature and Human Individuality Norton 2006 New York.

6. Dawkins, R.L., et al. Mutation versus polymorphism in evolution. Genomics 2013 101: 211-212.

7. Ayala, F.J. The Big Questions: Evolution. Ed. S. Blackburn, London: Quercus Publishing Plc, 2012.

8. Lewontin, R. C. The Genetic Basis of Evolutionary Change. New York: Columbia University Press, 1974.

9. Dawkins, R.L., et al. Genomics of the major histocompatibility complex: Haplotypes, duplication, retroviruses and disease. Immunol. Rev. 1999 167: 275-304.

10. Vandiedonck, C., et al. Pleiotropic effects of the 8.1 HLA haplotype in patients with autoimmune myasthenia gravis and thymus hyperplasia. Proc. Natl. Acad. Sci. USA 2004. 101:15464-15469.

11. Toth, E.K., et al. The 8.1 ancestral MHC haplotype is strongly associated with colorectal cancer risk. Intl. J. Cancer 2007 121: 1744–1748.

12. Lio, D., et al. Association between the MHC Class I gene HFE polymorphisms and longevity: A study in Sicilian population. Genes and Immunity 2002 3: 20-24.

13. Candore, G., et al. Pathogenesis of autoimmune disease associated with 8.1 ancestral haplotype: Effect of multiple gene interactions. Autoimmunity Rev. 2002 1: 29-35.

14. Vandiedonck, C., et al. Pervasive haplotypic variation in the spliceo-transcriptome of the MHC. Genome Res. 2011 s21: 1042-1054.

15. Medawar, P.B. *The Uniqueness of the Individual.* Dover, New York, 2nd ed 1981.

16. Burnet, F.M. *Immunological Surveillance*. Pergamon Press, Oxford 1970.

17. Burwell, R.G. The role of lymphoid tissue in morphostasis. Lancet 1963 ii, 69.

18. Sacks, O.W. Prosopagnosia: the man who mistook his wife for a hat and other clinical tales. Summit, New York 1985 and Farah 1992.

19. Cummins, H. and Midlo, C. Fingerprints, Palms and Soles: An Introduction to Dermatoglyphics. Dover, New York 1961.

20. Jain, Prabhakar and Pankani, Biometrics.cse.edu.au

21. Bodmer, W. and McKie, R. The Book of Man: The Human Genome Project and the Quest to Discover our Genetic Heritage. Oxford University Press, New York, 1994.

22. Burger, B., Fuchs, D., Sprecher, E. and Itin, P. The immigration delay disease: Adermatoglyphia-inherited absence of epidermal ridges. J. Am. Acad. Derm. 2011 64: 974-980.

23. Nousbeck, J., et al. A mutation in a skin-specific isoform of SMARCAD1 causes autosomal Dominant Adermatoglyphia. Am. J. Hum. Genet. 2011 89 (2) :302-307.

24. Wong, M. et al. Travel warning with Capecitabine. Ann. Oncology, 2009 20:1281-1286.

25. Szajerka, T., et al. Chimerism in hand transplantation. Ann. Transplant. 201116(1): 83-89.

26. Lenochova, P. and Havilack, J. Human Body Odour Individuality. In: J.L. Hurst et al. (Eds) Chemical Signals in Vertebrates. 11, Chapter 18 pp 189-198, 2008.

27. Hammond, C.J., et al. Genetic and environmental factors in age-related nuclear cataracts in monozygotic and dizygotic twins. New Engl. J. Med. 2000 342: 1786-1790.

28. Williamson, J.F., et al. Genomic evolution in domestic cattle: Ancestral haplotypes and healthy beef. Genomics 2011 97:304-312.

29. Laird, R., et al. Use of the genomic matching technique to complement multiplex STR profiling reduces DNA profiling costs in high volume crimes and intelligence led screens. Forensic Science International 2005 151 : 249–257.

****

Index

Made in the USA
Coppell, TX
26 November 2025

63993893R00107